J.O. Bird
B.Sc.(Hons), F.Coll.P., A.F.I.M.A.,
M.I.Elec.I.E.

A.J.C. May
B.A., C.Eng., M.I.Mech.E., F.I.T.E., A.M.B.I.M.

Calculus for technicians

Second edition

Longman London and New York

Longman Group Limited
Longman House, Burnt Mill, Harlow
Essex CM20 2JE, England
Associated companies throughout the world

Published in the United States of America
by Longman Inc., New York

First published 1979
Second edition 1985

British Library Cataloguing in Publication Data

Bird, J. O.
Calculus for technicians. – 2nd ed. –
(Longman technician series/Mathematics and sciences)
1. Calculus
I. Title II. May, A. J. C.
515'.0246 QA303

ISBN 0-582-41370-2

Produced by Longman Singapore Publishers (Pte) Ltd.
Printed in Singapore

Sector Editor – Mathematics and Sciences

D. R. Browning, B.Sc., F.R.S.C., C.Chem., A.R.T.C.S.
Formerly Principal Lecturer and Head of Chemistry, Bristol Polytechnic

Books already published in this sector of the series:

Technician mathematics, Level 1 **J. O. Bird and A. J. C. May**
Technician mathematics, Level 2 **J. O. Bird and A. J. C. May**
Technician mathematics, Level 3 **J. O. Bird and A. J. C. May**
Technician mathematics Levels 4 and 5 **J. O. Bird and A. J. C. May**
Mathematics for science technicians Level 2 **J. O. Bird and A. J. C. May**
Mathematics for electrical and telecommunications technicians Level 2 **J. O. Bird and A. J. C. May**
Mathematics for electrical technicians Level 3 **J. O. Bird and A. J. C. May**
Mathematics for electrical technicians Levels 4 and 5 **J. O. Bird and A. J. C. May**
Calculus for technicians **J. O. Bird and A. J. C. May**
Statistics for technicians **J. O. Bird and A. J. C. May**
Mathematical formulae for BTEC courses **J. O. Bird and A. J. C. May**
Algebra for technicians **J. O. Bird and A. J. C. May**
Safety science for technicians **W. J. Hackett and G. P. Robbins**
Fundamentals of chemistry **J. H. J. Peet**
Further studies in chemistry **J. H. J. Peet**
Technician chemistry, Level 1 **J. Brockington and P. J. Stamper**
Mathematics for scientific and technical students **H. G. Davies and G. A. Hicks**
Organic chemistry for higher education **J. Brockington and P. J. Stamper**
Inorganic chemistry for higher education **J. Brockington and P. J. Stamper**
Digital techniques Level 2 **D. R. Browning**
Microprocessors and control **J. F. A. Thompson**
Cell biology for technicians Level 2 **N. A. Thorpe**
Physical chemistry for higher education **J. Brockington, P. J. Stamper and D. R. Browning**
Algebra and calculus for technicians Level 3 **J. O. Bird and A. J. C. May**

Contents

Chapter 5 Applications of integration to areas, mean values and root mean square values 92

Chapter 6 Numerical integration 117

Chapter 7 Differential equations 135

Appendices 150

Preface

This textbook provides coverage of the Business and Technician Education Council Level 3 Calculus half unit. However it can be regarded as a basic textbook on the introduction to Calculus for a much wider range of courses.

The aim of the book is to enable the student to differentiate and integrate simple functions, to solve certain first-order differential equations and to apply calculus to scientific and technological problems.

Each topic considered in the text is presented in a way that assumes in the reader only the knowledge attained at BTEC Level 2 Mathematics.

This practical book contains some 50 illustrations, 120 detailed worked problems, followed by over 500 further problems with answers.

The authors would like to express their appreciation for the friendly cooperation and helpful advice given to them by the publishers and to thank Mrs. Elaine Woolley for the excellent typing of the manuscript.

Finally, the authors would like to add a word of thanks to their wives, Elizabeth and Juliet, for their patience, help and encouragement during the preparation of this book.

J. O. Bird
A. J. C. May

Highbury College of Technology
Portsmouth

Chapter 1

Introduction to differentiation

1 Introduction

Calculus is a branch of mathematics involving or leading to calculations dealing with continuously varying functions. The subject falls into two parts, namely **differential calculus** (usually abbreviated to **differentiation**) and **integral calculus** (usually abbreviated to **integration**).

The central problem of the differential calculus is the investigation of the rate of change of a function with respect to changes in the variables on which it depends.

The two main uses of integral calculus are firstly, finding such quantities as the length of a curve, the area enclosed by a curve, or the volume enclosed by a surface, and secondly, the problem of determining a variable quantity given its rate of change.

There is a close relationship between the processes of differentiation and integration, the latter being considered as the inverse of the former.

Calculus is a comparatively young branch of mathematics; its systematic development started in the middle of the 17th century. Since then there has been an enormous expansion in the scope of calculus and it is now used in every field of applied science as an instrument for the solution of problems of the most varied nature.

Before such uses can be investigated it is essential to grasp the basic concepts and to understand the notations used. The following text deals with this necessary preparatory work.

2 Functional notation

An expression such as $y = 4x^2 - 4x - 3$ contains two variables. For every value of x there is a corresponding value of y. The variable x is called the **independent variable** and y is called the **dependent variable.** y is said to be a function of x and is written as $y = f(x)$. Hence from above $f(x) = 4x^2 - 4x - 3$.

The value of the function $f(x)$ when $x = 0$ is denoted by $f(0)$. Similarly when $x = 1$ the value of the function is denoted by $f(1)$ and so on.

If $f(x) = 4x^2 - 4x - 3$

then $f(0) = 4(0)^2 - 4(0) - 3 = -3$

$f(1) = 4(1)^2 - 4(1) - 3 = -3$

$f(2) = 4(2)^2 - 4(2) - 3 = 5$

$f(3) = 4(3)^2 - 4(3) - 3 = 21$

$f(-1) = 4(-1)^2 - 4(-1) - 3 = 5$

and $f(-2) = 4(-2)^2 - 4(-2) - 3 = 21$

Figure 1 shows the curve $f(x) = 4x^2 - 4x - 3$ for values of x between $x = -2$ and $x = 3$. The lengths represented by $f(0)$, $f(1)$, $f(2)$, etc. are also shown.

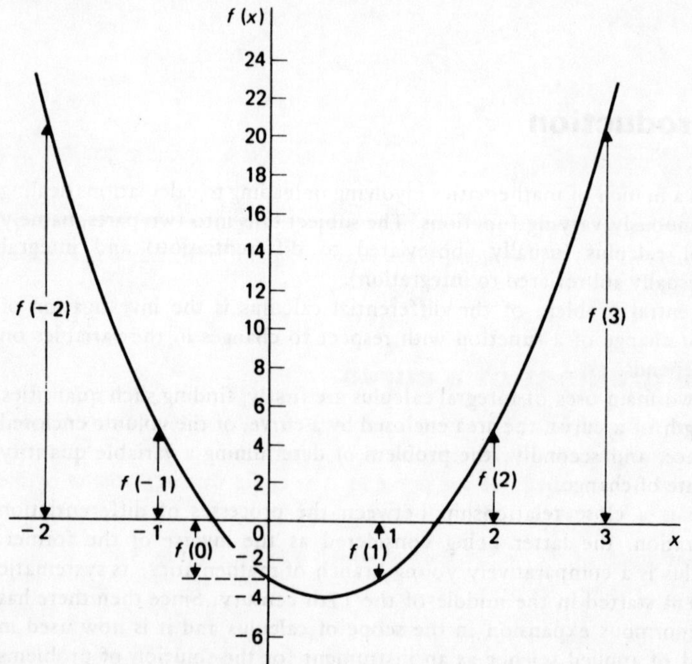

Fig. 1 Graph of $f(x) = 4x^2 - 4x - 3$

Problem 1. If $f(x) = 5x \ - 3x + 1$ find $f(0)$, $f(3)$, $f(-1)$, $f(-2)$ and $f(3) - f(-2)$.

$$\begin{aligned} f(x) &= 5x^2 - 3x + 1 & \\ f(0) &= 5(0)^2 - 3(0) + 1 &= 1 \\ f(3) &= 5(3)^2 - 3(3) + 1 &= 37 \\ f(-1) &= 5(-1)^2 - 3(-1) + 1 &= 9 \\ f(-2) &= 5(-2)^2 - 3(-2) + 1 &= 27 \\ f(3) - f(-2) &= 37 - 27 &= 10 \end{aligned}$$

Problem 2. For the curve $f(x) = 3x^2 + 2x - 9$ evaluate $f(2) \div f(1)$, $f(2 + a)$, $f(2 + a) - f(2)$ and $\dfrac{f(2 + a) - f(2)}{a}$.

$$\begin{aligned} f(x) &= 3x^2 + 2x - 9 \\ f(1) &= 3(1)^2 + 2(1) - 9 = -4 \\ f(2) &= 3(2)^2 + 2(2) - 9 = 7 \\ f(2 + a) &= 3(2 + a)^2 + 2(2 + a) - 9 \\ &= 3(4 + 4a + a^2) + 4 + 2a - 9 \\ &= 12 + 12a + 3a^2 + 4 + 2a - 9 \\ &= 7 + 14a + 3a^2 \end{aligned}$$

$$f(2) \div f(1) = \frac{f(2)}{f(1)} = \frac{7}{-4} = -1.$$

$$\begin{aligned} f(2 + a) - f(2) &= 7 + 14a + 3a^2 - 7 \\ &= 14a + 3a^2 \end{aligned}$$

$$\frac{f(2 + a) - f(2)}{a} = \frac{14a + 3a^2}{a} = 14 + 3a$$

Further problems on functional notation may be found in Section 5, problems 1 to 5, page 15.

3 The gradient of a curve

If a tangent is drawn at a point A on a curve then the gradient of this tangent is said to be the gradient of the curve at A.

In Fig. 2 the gradient of the curve at A is equal to the gradient of the tangent AB.

Consider the graph of $f(x) = 2x^2$, part of which is shown in Fig. 3.

The gradient of the chord PQ is given by:

$$\frac{QR}{PR} = \frac{QS - RS}{PR} = \frac{QS - PT}{PR}$$

At point P, $x = 1$ and at point Q, $x = 3$.

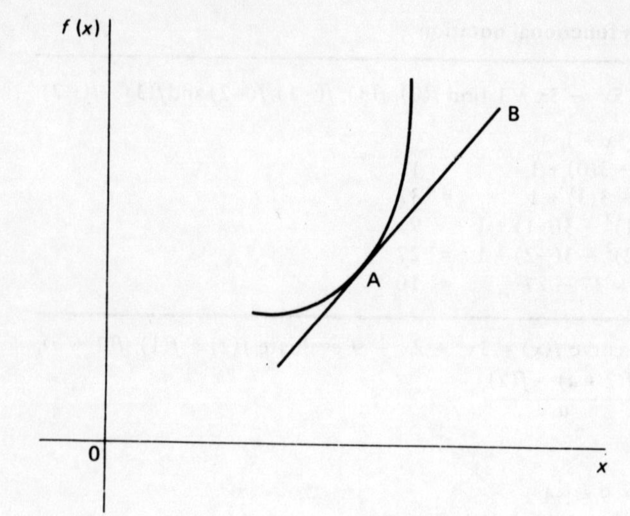

Fig. 2

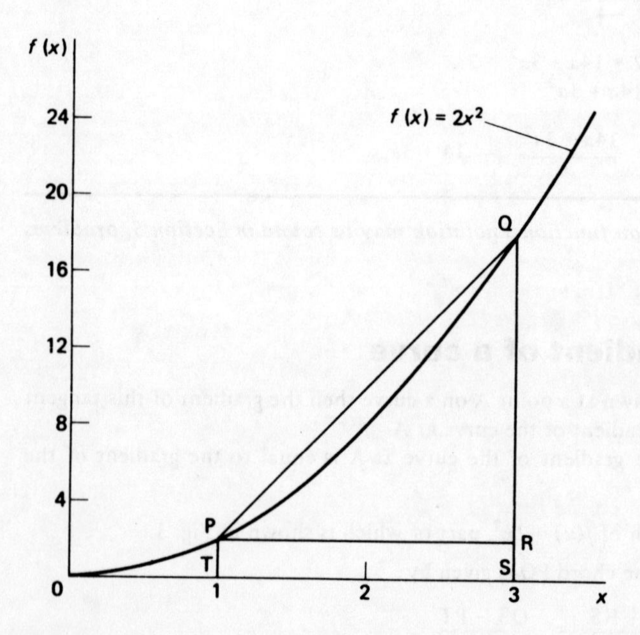

Fig. 3

Hence the gradient of the chord PQ $= \dfrac{f(3) - f(1)}{3 - 1}$

$$= \frac{18 - 2}{2}$$

$$= \frac{16}{2} = 8$$

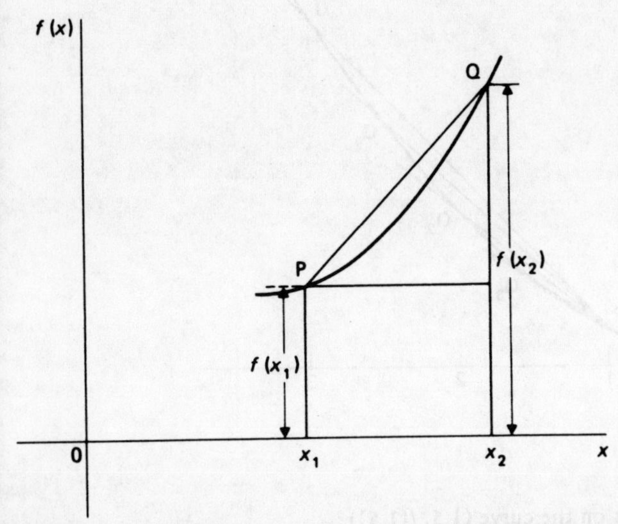

Fig. 4

More generally, for any curve (as shown in Fig. 4):

$$\text{Gradient of PQ} = \frac{f(x_2) - f(x_1)}{x_2 - x_1}$$

For the part of the curve $f(x) = 2x^2$ shown in Fig. 5 let us consider what happens as the point Q, at present at $(3, f(3))$, moves closer and closer to point P, which is fixed at $(1, f(1))$.

Let Q_1 be the point on the curve $(2.5, f(2.5))$.

$$\text{Gradient of chord } PQ_1 = \frac{f(2.5) - f(1)}{2.5 - 1}$$

$$= \frac{12.5 - 2}{1.5} = 7$$

Let Q_2 be the point on the curve $(2, f(2))$.

$$\text{Gradient of chord } PQ_2 = \frac{f(2) - f(1)}{2 - 1}$$

$$= \frac{8 - 2}{1} = 6$$

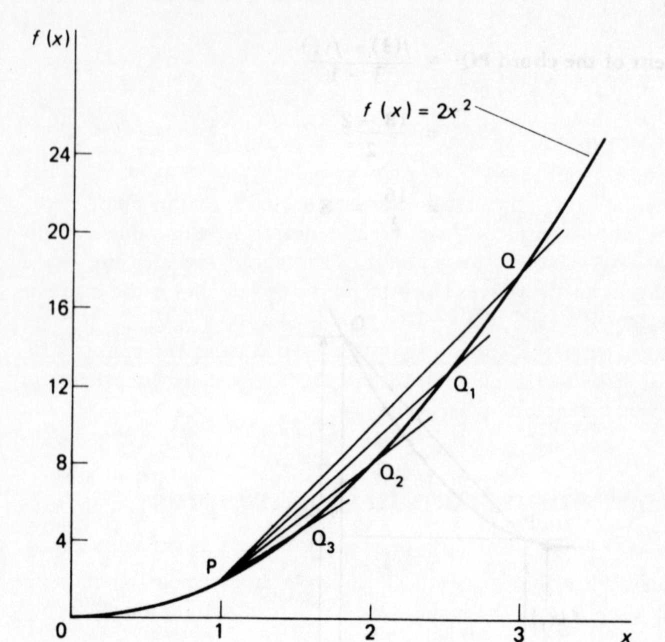

Fig. 5

Let Q_3 be the point on the curve $(1.5, f(1.5))$.

$$\text{Gradient of chord } PQ_3 = \frac{f(1.5) - f(1)}{1.5 - 1}$$

$$= \frac{4.5 - 2}{0.5} = 5$$

The following points, i.e. Q_4, Q_5 and Q_6, are not shown on Fig. 5.
Let Q_4 be the point on the curve $(1.1,\ f(1.1))$.

$$\text{Gradient of chord } PQ_4 = \frac{f(1.1) - f(1)}{1.1 - 1}$$

$$= \frac{2.42 - 2}{0.1} = 4.2$$

Let Q_5 be the point on the curve $(1.01, f(1.01))$.

$$\text{Gradient of chord } PQ_5 = \frac{f(1.01) - f(1)}{1.01 - 1}$$

$$= \frac{2.040\,2 - 2}{0.01} = 4.02$$

Let Q_6 be the point on the curve $(1.001, f(1.001))$.

$$\text{Gradient of chord } PQ_6 = \frac{f(1.001) - f(1)}{1.001 - 1}$$

$$= \frac{2.004\,002 - 2}{0.001} = \mathbf{4.002}$$

Thus as the point Q approaches closer and closer to the point P the gradients of the chords approach nearer and nearer to the value 4. This is called the **limiting value** of the gradient of the chord and **at P the chord becomes the tangent to the curve.** Thus the limiting value of 4 is the gradient of the tangent at P.

It can be seen from the above example that deducing the gradient of the tangent to a curve at a given point by this method is a lengthy process. A much more convenient method is shown below.

4 Differentiation from first principles

Let P and Q be two points very close together on a curve as shown in Fig. 6.

Let the length PR be δx (pronounced delta x), representing a small increment (or increase) in x, and the length QR, the corresponding increase in y,

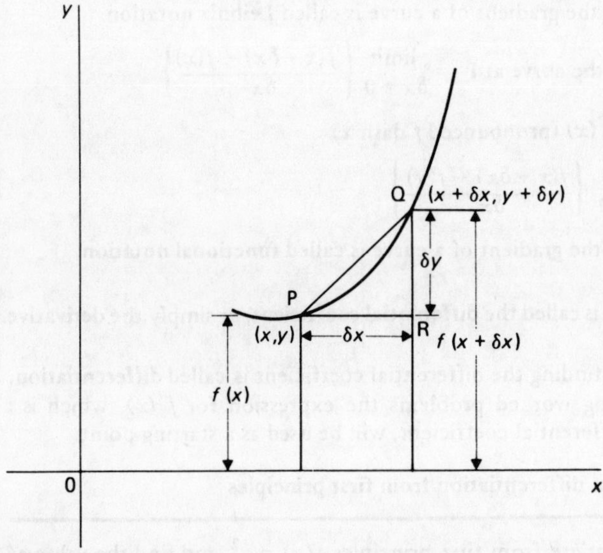

Fig. 6

be δy (pronounced delta y). It is important to realise that δ and x are inseparable, i.e. δx does not mean δ times x. Let P be any point on the curve with coordinates (x, y). Then Q will have the coordinates $(x + \delta x, y + \delta y)$.

$$\text{The slope of the chord PQ} = \frac{\delta y}{\delta x}$$

But from Fig. 6, $\delta y = (y + \delta y) - y = f(x + \delta x) - f(x)$

Hence $\dfrac{\delta y}{\delta x} = \dfrac{f(x + \delta x) - f(x)}{\delta x}$

The smaller δx becomes, the closer the gradient of the chord PQ approaches the gradient of the tangent at P. That is, as $\delta x \to 0$, the gradient of the chord $\to$ the gradient of the tangent. (Note '$\to$' means 'approaches'). As δx approaches zero, the value of $\dfrac{\delta y}{\delta x}$ approaches what is called a **limiting value.** There are two notations commonly used when finding the gradient of a tangent drawn to a curve.

1. The gradient of the curve at P is represented as $\underset{\delta x \to 0}{\text{limit}} \dfrac{\delta y}{\delta x}$

This is written as $\dfrac{dy}{dx}$ (pronounced dee y by dee x), i.e.

$$\frac{dy}{dx} = \underset{\delta x \to 0}{\lim.} \frac{\delta y}{\delta x}$$

This way of stating the gradient of a curve is called **Leibniz notation.**

2. The gradient of the curve at P $= \underset{\delta x \to 0}{\text{limit}} \left\{ \dfrac{f(x + \delta x) - f(x)}{\delta x} \right\}$

This is written as $f'(x)$ (pronounced f dash x)

i.e. $f'(x) = \underset{\delta x \to 0}{\text{limit}} \left\{ \dfrac{f(x + \delta x) - f(x)}{\delta x} \right\}$

This way of stating the gradient of a curve is called **functional notation.**

$\dfrac{dy}{dx}$ equals $f'(x)$ and is called the **differential coefficient,** or simply the **derivative.**

The process of finding the differential coefficient is called **differentiation.** In the following worked problems the expression for $f'(x)$, which is a definition of the differential coefficient, will be used as a starting point.

Worked problems on differentiation from first principles

Problem 1. Differentiate from first principles $f(x) = x^2$ and find the value of the gradient of the curve at $x = 3$.

To 'differentiate from first principles' means 'to find $f'(x)$' by using the expression:

$$f'(x) = \underset{\delta x \to 0}{\text{limit}} \left\{ \frac{f(x + \delta x) - f(x)}{\delta x} \right\}$$

$f(x) = x^2$

$f(x + \delta x) = (x + \delta x)^2 = x^2 + 2.x\,\delta x + \delta x^2$

$$f(x + \delta x) - f(x) = x^2 + 2\,x\,\delta x + \delta x^2 - x^2$$
$$= 2\,x\,\delta x + \delta x^2$$

$$\frac{f(x + \delta x) - f(x)}{\delta x} = \frac{2x\,\delta x + \delta x^2}{\delta x}$$
$$= 2x + \delta x$$

As $\delta x \to 0$, $\dfrac{f(x + \delta x) - f(x)}{\delta x} \to 2x + 0$

Therefore $f'(x) = \underset{\delta x \to 0}{\text{limit}} \left\{ \dfrac{f(x + \delta x) - f(x)}{\delta x} \right\} = 2x$

At $x = 3$, the gradient of the curve, i.e. $f'(x) = 2(3) = 6$

Hence if $f(x) = x^2$, $f'(x) = 2x$. The gradient at $x = 3$ is 6.

Problem 2. Find the differential coefficient of $f(x) = 3x^3$, from first principles.

To 'find the differential coefficient' means 'to find $f'(x)$' by using the expression:

$$f'(x) = \underset{\delta x \to 0}{\text{limit}} \left\{ \frac{f(x + \delta x) - f(x)}{\delta x} \right\}$$

$f(x) = 3x^3$

$$f(x + \delta x) = 3(x + \delta x)^3$$
$$= 3(x + \delta x)\,(x^2 + 2\,x\,\delta x + \delta x^2)$$
$$= 3(x^3 + 3x^2\delta x + 3x\delta x^2 + \delta x^3)$$
$$= 3x^3 + 9x^2\delta x + 9x\delta x^2 + 3\delta x^3.$$

$$f(x + \delta x) - f(x) = 3x^3 + 9x^2\delta x + 9x\delta x^2 + 3\delta x^3 - 3x^3$$
$$= 9x^2\,\delta x + 9x\delta x^2 + 3\delta x^3$$

$$\frac{f(x + \delta x) - f(x)}{\delta x} = \frac{9x^2\delta x + 9x\delta x^2 + 3\delta x^3}{\delta x}$$
$$= 9x^2 + 9x\delta x + 3\delta x^2$$

As $\delta x \to 0$, $\dfrac{f(x + \delta x) - f(x)}{\delta x} \to 9x^2 + 9x(0) + 3(0)^2$

i.e. $f'(x) = \underset{\delta x \to 0}{\text{limit}} \left\{ \dfrac{f(x + \delta x) - f(x)}{\delta x} \right\} = 9x^2$

Problem 3. By differentiation from first principles determine $\dfrac{dy}{dx}$ for $y = 3x$.

The object is to find $\dfrac{dy}{dx}$

$$\frac{dy}{dx} = f'(x) = \underset{\delta x \to 0}{\text{limit}} \left\{ \frac{f(x + \delta x) - f(x)}{\delta x} \right\}$$

$$y = f(x) = 3x$$

$$f(x + \delta x) = 3(x + \delta x) = 3x + 3\delta x$$

$$f(x + \delta x) - f(x) = 3x + 3\delta x - 3x$$

$$= 3\delta x$$

$$\frac{f(x + \delta x) - f(x)}{\delta x} = \frac{3\delta x}{\delta x} = 3$$

Hence $\dfrac{dy}{dx} = \underset{\delta x \to 0}{\text{limit}} \left\{ \dfrac{f(x + \delta x) - f(x)}{\delta x} \right\} = 3.$

Another way of writing $\dfrac{dy}{dx} = 3$ is $f'(x) = 3$

or $\dfrac{d}{dx}(3x) = 3$ since $y = 3x$.

Problem 4. Find the derivative of $y = \sqrt{x}$.

Let $y = f(x) = \sqrt{x}$
To 'find the derivative' means 'to find $f'(x)$'.

$$f'(x) = \underset{\delta x \to 0}{\text{lim.}} \left\{ \frac{f(x + \delta x) - f(x)}{\delta x} \right\}$$

$$f(x) = \sqrt{x} = x^{\frac{1}{2}}$$

$$f(x + \delta x) = (x + \delta x)^{\frac{1}{2}}$$

$$f(x + \delta x) - f(x) = (x + \delta x)^{\frac{1}{2}} - x^{\frac{1}{2}}$$

$$\frac{f(x + \delta x) - f(x)}{\delta x} = \frac{(x + \delta x)^{\frac{1}{2}} - x^{\frac{1}{2}}}{\delta x}$$

Now from algebra, $(a - b)(a + b) = a^2 - b^2$, i.e. the difference of two squares. Therefore, in this case, multiplying both the numerator and the denominator by $[(x + \delta x)^{\frac{1}{2}} + x^{\frac{1}{2}}]$, to make the numerator of the fraction of $(a+b)(a-b)$ form, gives:

$$\frac{f(x + \delta x) - f(x)}{\delta x} = \frac{[(x + \delta x)^{\frac{1}{2}} - x^{\frac{1}{2}}]\,[(x + \delta x)^{\frac{1}{2}} + x^{\frac{1}{2}}]}{\delta x\,[(x + \delta x)^{\frac{1}{2}} + x^{\frac{1}{2}}]}$$

$$= \frac{[(x + \delta x)^{\frac{1}{2}}]^2 - [x^{\frac{1}{2}}]^2}{\delta x\,[(x + \delta x)^{\frac{1}{2}} + x^{\frac{1}{2}}]}$$

$$= \frac{(x + \delta x) - (x)}{\delta x\,[(x + \delta x)^{\frac{1}{2}} + x^{\frac{1}{2}}]}$$

$$= \frac{\delta x}{\delta x\ [(x+\delta x)^{\frac{1}{2}} + x^{\frac{1}{2}}]}$$

$$= \frac{1}{(x+\delta x)^{\frac{1}{2}} + x^{\frac{1}{2}}}$$

As $\delta x \to 0$, $\dfrac{f(x+\delta x) - f(x)}{\delta x} \to \dfrac{1}{(x+0)^{\frac{1}{2}} + x^{\frac{1}{2}}}$

Therefore $f'(x) = \lim\limits_{\delta x \to 0} \left\{ \dfrac{f(x+\delta x) - f(x)}{\delta x} \right\} = \dfrac{1}{x^{\frac{1}{2}} + x^{\frac{1}{2}}} = \dfrac{1}{2\,x^{\frac{1}{2}}}$

Hence if $f(x) = \sqrt{x}$, $f'(x) = \dfrac{1}{2\sqrt{x}}$ or $\frac{1}{2}x^{-\frac{1}{2}}$

Another way of writing this is:

If $y = \sqrt{x}$, $\dfrac{dy}{dx} = \dfrac{1}{2\sqrt{x}}$,

or $\dfrac{d}{dx}(\sqrt{x}) = \dfrac{1}{2\sqrt{x}}$

Problem 5. Differentiate from first principles $f(x) = \dfrac{1}{2x}$

$$f'(x) = \lim_{\delta x \to 0} \left\{ \frac{f(x+\delta x) - f(x)}{\delta x} \right\}$$

$$f(x) = \frac{1}{2x}$$

$$f(x+\delta x) = \frac{1}{2(x+\delta x)}$$

$$f(x+\delta x) - f(x) = \frac{1}{2(x+\delta x)} - \frac{1}{2x}$$

$$= \frac{x - (x+\delta x)}{2x\ (x+\delta x)}$$

$$= \frac{-\delta x}{2x\ (x+\delta x)}$$

$$\frac{f(x+\delta x) - f(x)}{\delta x} = \frac{-\delta x}{2x\ (x+\delta x)\ \delta x}$$

$$= \frac{-1}{2x\ (x+\delta x)}$$

As $\delta x \to 0$, $\dfrac{f(x+\delta x)-f(x)}{\delta x} \to \dfrac{-1}{2x\,(x+0)}$

Therefore $f'(x) = \lim\limits_{\delta x \to 0}\left\{\dfrac{f(x+\delta x)-f(x)}{\delta x}\right\} = \dfrac{-1}{2x\,(x)} = -\dfrac{1}{2x^2}$

Problem 6. Find the differential coefficient of $y = 5$.

The differential coefficient of $y = 5$ may be deduced as follows:
If a graph is drawn of $y = 5$ a straight horizontal line results and the gradient or slope of a horizontal line is zero. Finding the differential coefficient is, in fact, finding the slope of a curve, or, as in this case, of a horizontal straight line.

Hence $\dfrac{dy}{dx} = 0$

This may also be shown by the conventional method since:

$$\frac{dy}{dx} = f'(x) = \lim_{\delta x \to 0}\left\{\frac{f(x+\delta x)-f(x)}{\delta x}\right\}$$

$y = f(x) = 5$
$\therefore\ f(x+\delta x) = 5$

$$\therefore\ \frac{dy}{dx} = f'(x) = \lim_{\delta x \to 0}\left\{\frac{5-5}{\delta x}\right\}$$

$$= \frac{0}{\delta x} = 0$$

More generally, if C is any constant, then if
$f(x) = C,\ f'(x) = 0$

i.e. If $y = C$ then $\dfrac{dy}{dx} = 0$

Problem 7. Differentiate from first principles $f(x) = 3x^2 + 6x - 3$ and find the gradient of the curve at $x = -2$.

$$f'(x) = \lim_{\delta x \to 0}\left\{\frac{f(x+\delta x)-f(x)}{\delta x}\right\}$$

$f(x) = 3x^2 + 6x - 3$
$\therefore f(x+\delta x) = 3(x+\delta x)^2 + 6(x+\delta x) - 3$
$= 3(x^2 + 2x\,\delta x + \delta x^2) + 6x + 6\delta x - 3$
$= 3x^2 + 6x\delta x + 3\delta x^2 + 6x + 6\delta x - 3$
$\therefore f(x+\delta x) - f(x) = (3x^2 + 6x\delta x + 3\delta x^2 + 6x + 6\delta x - 3) - (3x^2 + 6x - 3)$
$= 6x\,\delta x + 3\delta x^2 + 6\delta x$

$$\therefore\ \frac{f(x+\delta x)-f(x)}{\delta x} = \frac{6x\,\delta x + 3\delta x^2 + 6\delta x}{\delta x}$$

$$= 6x + 3\delta x + 6$$

As $\delta x \to 0$, $\dfrac{f(x + \delta x) - f(x)}{\delta x} \to 6x + 3\,(0) + 6$

Therefore $f'(x) = \lim\limits_{\delta x \to 0} \left\{ \dfrac{f(x + \delta x) - f(x)}{\delta x} \right\} = 6x + 6$

At $x = -2$ the gradient of the curve, i.e. $f'(x)$, is $6(-2) + 6$, i.e. -6.
Hence if $f(x) = 3x^2 + 6x - 3$, $f'(x) = 6x + 6$ and the gradient of the curve at $x = -2$ is -6.

A summary of the results obtained in the above problems is tabulated below:

y or $f(x)$	$\dfrac{dy}{dx}$ or $f'(x)$
x^2	$2x$
$3x^3$	$9x^2$
$3x$	3
$x^{\frac{1}{2}}$	$\frac{1}{2}x^{-\frac{1}{2}}$
$\dfrac{1}{2x}$	$-\dfrac{1}{2x^2}$
5	0
$3x^2 + 6x - 3$	$6x + 6$

Three basic rules of differentiation emerge from these results:
Rule 1. The differential coefficient of a constant is zero.

Rule 2. $\dfrac{d}{dx}(x^n) = nx^{n-1}$.

For example $\dfrac{d}{dx}(x^3) = 3x^{3-1} = 3x^2$ (as in the table)

Rule 3. Constants associated with variables are carried forward.

For example $\dfrac{d}{dx}(3x^2) = 3\dfrac{d}{dx}(x^2)$

Problem 8. Differentiate from first principles $f(x) = \dfrac{1}{5x + 3}$

$$f'(x) = \lim_{\delta x \to 0} \left\{ \frac{f(x + \delta x) - f(x)}{\delta x} \right\}$$

$$f(x) = \frac{1}{5x + 3}$$

$$f(x + \delta x) = \frac{1}{5(x + \delta x) + 3}$$

$$f(x + \delta x) - f(x) = \frac{1}{5(x + \delta x) + 3} - \frac{1}{5x + 3}$$

$$= \frac{[5x + 3] - [5(x + \delta x) + 3]}{[5\,(x + \delta x) + 3]\ [5x + 3]}$$

$$= \frac{5x + 3 - 5x - 5\delta x - 3}{[5(x + \delta x) + 3]\ [5x + 3]}$$

$$= \frac{-5\,\delta x}{[5(x + \delta x) + 3]\ [5x + 3]}$$

$$\therefore \frac{f(x + \delta x) - f(x)}{\delta x} = \frac{-5\,\delta x}{[5(x + \delta x) + 3]\ [5x + 3]\ \delta x}$$

$$= \frac{-5}{[5(x + \delta x) + 3]\ [5x + 3]}$$

As $\delta x \to 0$, $\dfrac{f(x + \delta x) - f(x)}{\delta x} \to \dfrac{-5}{[5(x + 0) + 3]\ [5x + 3]}$

Therefore $f'(x) = \lim_{\delta x \to 0} \left\{ \dfrac{f(x + \delta x) - f(x)}{\delta x} \right\} = \dfrac{-5}{(5x + 3)\,(5x + 3)}$

Hence if $f(x) = \dfrac{1}{5x + 3}$, $f'(x) = \dfrac{-5}{(5x + 3)^2}$

In the above worked problems the questions have been worded in a variety of ways. The important thing to realise is that they all mean the same thing. For example, in worked problem 8, on differentiating from first principles

$f(x) = \dfrac{1}{5x + 3}$ gives $\dfrac{-5}{(5x + 3)^2}$. This result can be expressed in a number of ways.

1. If $f(x) = \dfrac{1}{5x + 3}$ then $f'(x) = \dfrac{-5}{(5x + 3)^2}$

2. If $y = \dfrac{1}{5x + 3}$ then $\dfrac{dy}{dx} = \dfrac{-5}{(5x + 3)^2}$

3. The differential coefficient of $\dfrac{1}{5x + 3}$ is $\dfrac{-5}{(5x + 3)^2}$

4. The derivative of $\dfrac{1}{5x + 3}$ is $\dfrac{-5}{(5x + 3)^2}$

5. $\dfrac{d}{dx}\left(\dfrac{1}{5x + 3}\right) = \dfrac{-5}{(5x + 3)^2}$

Further problems on differentiating from first principles may be found in the following Section (numbers 6 to 34).

5 Further problems

Functional notation

1. If $f(x) = 2x^2 - x + 3$ find $f(0)$, $f(1)$, $f(2)$, $f(-1)$ and $f(-2)$.
[3, 4, 9, 6, 13]

2. If $f(x) = 6x^2 - 4x + 7$ find $f(1)$, $f(2)$, $f(-2)$ and $f(1) - f(-2)$.
[9, 23, 39, −30]

3. If a curve is represented by $f(x) = 2x^3 + x^2 - x + 6$ prove that $f(1) = \frac{1}{3}f(2)$.

4. If $f(x) = 3x^2 + 2x - 9$ find $f(3)$, $f(3 + a)$ and $\dfrac{f(3+a) - f(3)}{a}$.
[24, $3a^2 + 20a + 24$, $3a + 20$]

5. If $f(x) = 4x^3 - 2x^2 - 3x + 1$ find $f(2)$, $f(-3)$ and $\dfrac{f(1 + b) - f(1)}{b}$.
[19, -116, $4b^2 + 10b + 5$]

Differentiation from first principles

6. Sketch the curve $f(x) = 5x^2 - 6$ for values of x from $x = -2$ to $x = +4$. Label the coordinate $(3.5, f(3.5))$ as A. Label the coordinate $(1.5, f(1.5))$ as B. Join points A and B to form the chord AB. Find the gradient of the chord AB. By moving A nearer and nearer to B find the gradient of the tangent of the curve at B. [25, 15]

In problems 7–27 differentiate from first principles:

7. $y = x$. [1]
8. $y = 5x$. [5]
9. $y = x^2$. [$2x$]
10. $y = 7x^2$. [$14x$]
11. $y = 4x^3$. [$12x^2$]
12. $y = 2x^2 - 3x + 2$. [$4x - 3$]
13. $y = 2\sqrt{x}$. $\left[\dfrac{1}{\sqrt{x}} \text{ or } x^{-\frac{1}{2}}\right]$
14. $y = \dfrac{1}{x}$. $\left[-\dfrac{1}{x^2}\right]$
15. $y = \dfrac{5}{6x^2}$. $\left[-\dfrac{5}{3x^3}\right]$
16. $y = 19$. [0]

17. $f(x) = 3x$. [3]

18. $f(x) = \frac{x}{4}$. $\left[\frac{1}{4}\right]$

19. $f(x) = 3x^2$. [$6x$]

20. $f(x) = 14x^3$. [$42x^2$]

21. $f(x) = x^2 + 16x - 4$. [$2x + 16$]

22. $f(x) = 4x^{\frac{1}{2}}$. $\left[2x^{-\frac{1}{2}} \text{ or } \frac{2}{\sqrt{x}}\right]$

23. $f(x) = \frac{16}{17x}$. $\left[-\frac{16}{17x^2}\right]$

24. $f(x) = \frac{1}{x^3}$. $\left[-\frac{3}{x^4}\right]$

25. $f(x) = 8$. [0]

26. $f(x) = \frac{1}{\sqrt{x}}$ $\left[-\frac{1}{2\sqrt{x^3}} \text{ or } -\frac{1}{2}x^{-3/2}\right]$

27. $f(x) = \frac{1}{3x-2}$ $\left[\frac{-3}{(3x-2)^2}\right]$

28. Find $\frac{d}{dx}(6x^3)$. [$18x^2$]

29. Find $\frac{d}{dx}(3\sqrt{x} + 6)$ $\left[\frac{3}{2\sqrt{x}}\right]$

30. Find $\frac{d}{dx}(2x^{-2} + 7x^2)$. [$-4x^{-3} + 14x$]

31. Find $\frac{d}{dx}\left(13 - \frac{3}{2x}\right)$. $\left[\frac{3}{2x^2}\right]$

32. If E, F and G are the points (1, 2), (2, 16) and (3, 54) respectively on the graph of $y = 2x^3$, find the gradients of the tangents at the points E, F and G and the gradient of the chord EG. [6, 24, 54, 26]

33. Differentiate from first principles $f(x) = 5x^2 - 6x + 2$ and find the gradient of the curve at $x = 2$. [$10x - 6$, 14]

34. If $y = \frac{7}{2}\sqrt{x} + \frac{3}{x^2} - 9$ find the differential coefficient of y with respect to x.

$\left[\frac{7}{4\sqrt{x}} - \frac{6}{x^3}\right]$

Chapter 2

Methods of differentiation

1 Differential coefficients of some mathematical functions

(i) Differential coefficient of ax^n.

In the worked problems of Chapter 1 the differential coefficients of certain algebraic functions of the form $y = ax^n$ are derived from first principles and the results are summarised on page 13. The rules stated on page 13 are best remembered by the single statement that

$$\text{when } y = ax^n, \frac{dy}{dx} = anx^{n-1}$$

An analytical proof of this rule is given in Appendix D, (page 153).

(ii) Differential coefficient of sin x

A graph of $y = \sin x$ is shown in Fig. 1(a). The slope or gradient of the curve at any point is given by $\frac{dy}{dx}$ and is continually changing as values of x vary from O to S. By drawing tangents to the curve at many points on the curve

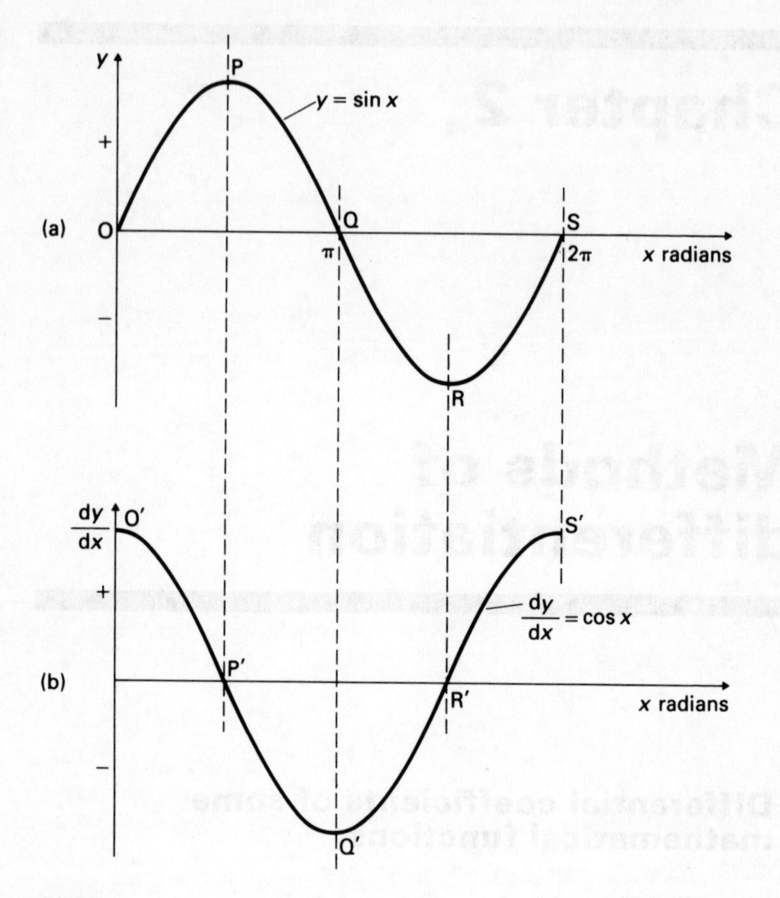

Fig. 1

and measuring the gradient of the tangents, values of $\frac{dy}{dx}$ may be obtained for corresponding values of x and these values are shown graphically in Fig. 1(b). The graph of $\frac{dy}{dx}$ against x so produced (called the derived curve) is a graph of $y = \cos x$. It follows that

$$\textbf{when } y = \sin x, \frac{dy}{dx} = \cos x$$

An analytical proof of this rule is given in Appendix D (page 153). By applying the principles of differentiation by substitution (see section 3 of this chapter), it may also be proved that

$$\textbf{when } y = \sin ax, \frac{dy}{dx} = a \cos ax$$

[An alternative method of reasoning the shape of the derived curve of $y = \sin x$ is as follows. By examining the curve of $y = \sin x$ in Fig. 1(a), the following observations can be made:

(i) at point O, the gradient is positive and at its steepest, giving a maximum positive value, shown by O′ in Fig. 1(b),
(ii) between O and P, values of the gradient are positive and decreasing in value, as values of x approach P,
(iii) at point P, the tangent is a horizontal line, hence the gradient is zero, shown as P′ in Fig. 1(b),
(iv) between P and Q, the gradient is negative and increasing in numerical value as x approaches point Q,
(v) at point Q the gradient is negative and at its steepest, giving a maximum negative value, shown by Q′ in Fig. 1(b).

Similarly, points R′ and S′ may be reasoned out for the negative half cycle of the curve $y = \sin x$.]

(iii) Differential coefficient of cos x

When graphs of $y = \cos x$ and its derived curve ($\frac{dy}{dx}$ against x) are drawn in a similar way to those for $y = \sin x$ shown in (ii) above, the derived curve is a graph of $(-\sin x)$. Thus

$$\textbf{when } y = \cos x, \ \frac{dy}{dx} = -\sin x$$

An analytical proof of this rule is given in Appendix D (page 154). By applying the principles of differentiation by substitution (see section 3 of this chapter), it may be proved that

$$\textbf{when } y = \cos ax, \ \frac{dy}{dx} = -a \sin ax$$

(iv) Differential coefficient of e^x

A graph of $y = e^x$ is shown in Fig. 2(a). The slope or gradient of the curve at any point is given by $\frac{dy}{dx}$ and is continually changing. By drawing tangents to the curve at many points on the curve and measuring the gradient of the tangents, values of $\frac{dy}{dx}$ for corresponding values of x may be obtained. These values are shown graphically in Fig. 2(b). The graph of $\frac{dy}{dx}$ against x so produced is identical to the original graph of $y = e^x$. It follows that

$$\textbf{when } y = e^x, \ \frac{dy}{dx} = e^x$$

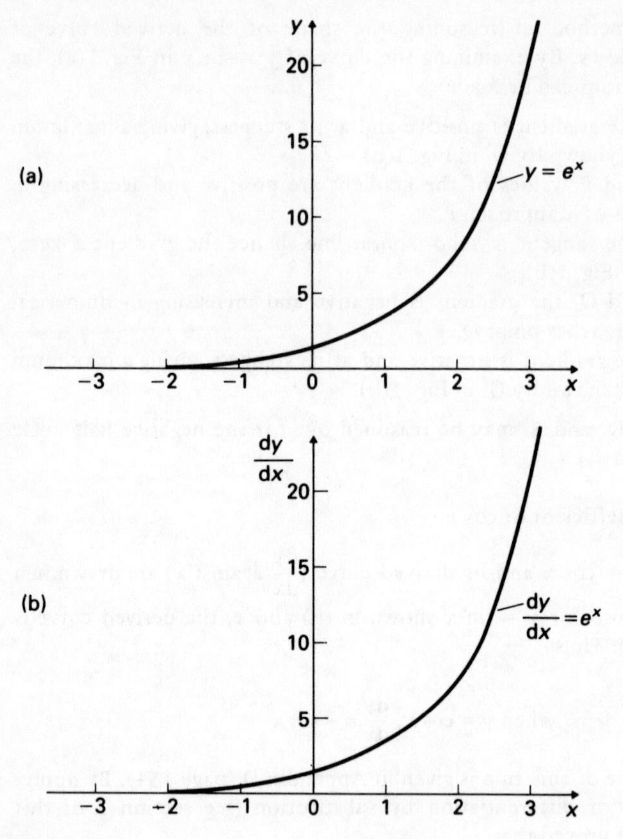

Fig. 2

By applying the principles of differentiation by substitution (see section 3 of this chapter), it may be proved that

$$\text{when } y = e^{ax}, \frac{dy}{dx} = a\,e^{ax}$$

This is as expected since by definition, the exponential function e^x, is a function whose rate of change is proportional to the original function. In the case of $y = e^x$, $a = 1$.

An analytical proof of this rule is given in Appendix D (page 155).

(v) Differential coefficient of ln x.

A graph of $y = \ln x$ is shown in Fig. 3(a). The slope or gradient of the curve at any point is given by $\frac{dy}{dx}$ and is continually changing. By drawing tangents to the curve at many points on the curve and measuring the slope of the

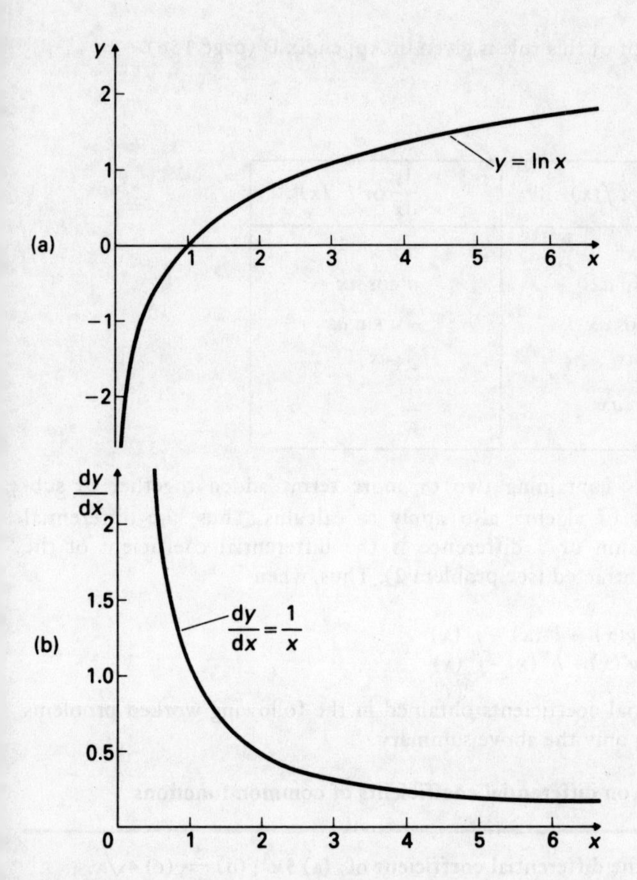

Fig. 3

tangents, values of $\frac{dy}{dx}$ for corresponding values of x may be obtained. These values are shown graphically in Fig. 3(b). The graph of $\frac{dy}{dx}$ against x so produced is the graph of $\frac{dy}{dx} = \frac{1}{x}$. It follows that

$$\text{when } y = \ln x, \frac{dy}{dx} = \frac{1}{x}$$

By applying the principles of differentiation by substitution (see section 3 of this chapter), it may be proved that

$$\text{when } y = \ln ax, \frac{dy}{dx} = \frac{1}{x}$$

(Note that when $y = \ln x$, $\frac{dy}{dx} \neq \frac{1}{ax}$)

 An analytical proof of this rule is given in Appendix D (page 156).

Summary

y or $f(x)$	$\frac{dy}{dx}$ or $f'(x)$
(i) ax^n	$a\,n\,x^{n-1}$
(ii) $\sin ax$	$a\cos ax$
(iii) $\cos ax$	$-a\sin ax$
(iv) e^{ax}	$a\,e^{ax}$
(v) $\ln ax$	$\frac{1}{x}$

For functions containing two or more terms added together or subtracted, the rules of algebra also apply to calculus. Thus, the differential coefficient of a sum or a difference is the differential coefficient of the terms added or subtracted (see problem 2). Thus, when

$$f(x) = g(x) + h(x) - j(x)$$
$$\text{then } f'(x) = g'(x) + h'(x) - j'(x)$$

The differential coefficients obtained in the following worked problems are deduced using only the above summary.

Worked problems on differential coefficients of common functions

---

Problem 1. Find the differential coefficient of: (a) $5x^4$; (b) $\frac{3}{x^2}$; (c) $4\sqrt{x}$.

When $f(x) = ax^n, f'(x) = an\,x^{n-1}$

(a) $f(x) = 5x^4$

$f'(x) = (5)(4)\,x^{4-1} = 20x^3$

(b) $f(x) = \frac{3}{x^2} = 3x^{-2}$

$f'(x) = (3)(-2)\,x^{-2-1} = -6x^{-3}$

i.e. $f'(x) = \frac{-6}{x^3}$

(c) $f(x) = 4\sqrt{x} = 4x^{\frac{1}{2}}$

$f'(x) = (4)(\tfrac{1}{2})\,x^{\frac{1}{2}-1} = 2x^{-\frac{1}{2}}$

i.e. $f'(x) = \frac{2}{\sqrt{x}}$

Problem 2. Differentiate $2x^3 + 7x + \frac{1}{3x^2} - \frac{4}{x^3} + \frac{4}{3}\sqrt{x^3} - 8$ with respect to x.

$$f(x) = 2x^3 + 7x + \frac{1}{3x^2} - \frac{4}{x^3} + \frac{4}{3}\sqrt{x^3} - 8$$

$$= 2x^3 + 7x + \frac{x^{-2}}{3} - 4x^{-3} + \frac{4}{3}x^{\frac{3}{2}} - 8$$

$$f'(x) = (2)(3)\,x^{3-1} + (7)\,x^{1-1} + \frac{(-2)}{3}x^{-2-1} - (4)(-3)\,x^{-3-1}$$

$$+ \left(\frac{4}{3}\right)\left(\frac{3}{2}\right)x^{\frac{3}{2}-1} - 0$$

$$= 6x^2 + 7 - \frac{2}{3}x^{-3} + 12\,x^{-4} + 2x^{\frac{1}{2}}$$

i.e. $f'(x) = 6x^2 + 7 - \dfrac{2}{3x^3} + \dfrac{12}{x^4} + 2\sqrt{x}$

Problem 3. (a) If $f(x) = 2 \sin x$ find $f'(x)$;

(b) If $y = 5 \cos x$ find $\dfrac{dy}{dx}$.

(a) When $f(x) = \sin x$, $f'(x) = \cos x$
When $f(x) = 2 \sin x$
then $f'(x) = 2 \cos x$

(b) When $y = \cos x$, $\dfrac{dy}{dx} = -\sin x$

When $y = 5 \cos x$

then $\dfrac{dy}{dx} = -5 \sin x$

Problem 4. Differentiate: (a) e^{6t}; (b) $5e^{-3t}$ with respect to t.

When $f(t) = e^{at}$, $f'(t) = a\,e^{at}$

(a) $f(t) = e^{6t}$
$f'(t) = 6\,e^{6t}$

(b) $f(t) = 5e^{-3t}$
$f'(t) = (5)(-3)\,e^{-3t} = -15\,e^{-3t}$

Problem 5. Find the differential coefficient of: (a) $\ln 4x$; (b) $3 \ln 2x$.

When $f(x) = \ln ax$, $f'(x) = \dfrac{1}{x}$

(a) $f(x) = \ln 4x$, $f'(x) = \dfrac{1}{x}$

(b) $f(x) = 3 \ln 2x$, $f'(x) = \dfrac{3}{x}$

Problem 6. If $g = 3.2\,h^5 - 3\sin h - 5e^{7h} + \sqrt[3]{h^4} + 6$, find $\dfrac{dg}{dh}$.

$$g = 3.2\,h^5 - 3\sin h - 5e^{7h} + h^{\frac{4}{3}} + 6$$

$$\frac{dg}{dh} = (3.2)(5)\,h^4 - 3\cos h - (5)(7)\,e^{7h} + \left(\frac{4}{3}\right) h^{\frac{1}{3}} + 0$$

$$= 16\,h^4 - 3\cos h - 35\,e^{7h} + \frac{4}{3}\,h^{\frac{1}{3}}$$

i.e. $$\frac{dg}{dh} = 16\,h^4 - 3\cos h - 35\,e^{7h} + \frac{4}{3}\sqrt[3]{h}$$

Problem 7. $f(x) = 4\ln(2.6x) - \dfrac{3}{\sqrt[3]{x^2}} + \dfrac{4}{e^{3x}} + \dfrac{1}{5} - 2\cos x$. Find $f'(x)$.

$$f(x) = 4\ln(2.6x) - 3x^{-\frac{2}{3}} + 4e^{-3x} + \frac{1}{5} - 2\cos x$$

$$f'(x) = \frac{4}{x} - (3)\left(-\frac{2}{3}\right)x^{-\frac{5}{3}} + (4)(-3)\,e^{-3x} + 0 - (-2\sin x)$$

i.e. $$f'(x) = \frac{4}{x} + \frac{2}{\sqrt[3]{x^5}} - \frac{12}{e^{3x}} + 2\sin x$$

Problem 8. Find the gradient of the curve $y = \dfrac{3}{2\sqrt{x}}$ at the point $\left(4, \dfrac{3}{4}\right)$.

$$y = \frac{3}{2\sqrt{x}} = \frac{3}{2}\,x^{-\frac{1}{2}}$$

$$\text{Gradient} = \frac{dy}{dx} = \left(\frac{3}{2}\right)\left(-\frac{1}{2}\right)x^{-\frac{3}{2}} = -\frac{3}{4\sqrt{x^3}}$$

When $x = 4$, gradient $= -\dfrac{3}{4\sqrt{4^3}} = -\dfrac{3}{4(8)}$

i.e. **Gradient** $= -\dfrac{3}{32}$

Problem 9. Find the coordinates of the point on the curve $y = 3\sqrt[3]{x^2}$ where the gradient is 1.

$$y = 3\sqrt[3]{x^2} = 3x^{\frac{2}{3}}$$

$$\text{Gradient} = \frac{dy}{dx} = (3)(\tfrac{2}{3})\,x^{-\frac{1}{3}} = \frac{2}{\sqrt[3]{x}}$$

If the gradient is equal to 1, then $1 = \dfrac{2}{\sqrt[3]{x}}$

i.e. $\sqrt[3]{x} = 2$

$x = 2^3 = 8$

When $x = 8$, $y = 3\sqrt[3]{8^2} = 3(4) = 12$

Hence the gradient is 1 at the point (8, 12)

Problem 10. If $f(x) = \dfrac{4x^3 - 8x^2 + 6x}{2x}$ find the coordinates of the point at which the gradient is: (a) zero; and (b) 4.

$$f(x) = \frac{4x^3 - 8x^2 + 6x}{2x} = \frac{4x^3}{2x} - \frac{8x^2}{2x} + \frac{6x}{2x}$$
$$= 2x^2 - 4x + 3$$

The derivative, $f'(x)$, gives the gradient of the curve.

Hence $f'(x) = (2)(2)\,x^1 - 4$
$= 4x - 4$

(a) When $f'(x)$ is zero
$4x - 4 = 0$
i.e. $x = 1$

When $x = 1$, $y = \dfrac{4x^3 - 8x^2 + 6x}{2x} = \dfrac{4(1)^3 - 8(1)^2 + 6(1)}{2(1)}$

i.e. $y = 1$

Hence the coordinates of the point where the gradient is zero are (1, 1)

(b) When $f'(x)$ is 4
$4x - 4 = 4$
i.e. $x = 2$

When $x = 2$, $y = 2\,(2)^2 - 4\,(2) + 3 = 3$

Hence the coordinates of the point where the gradient is 4 are (2, 3)

Further problems on differential coefficients of common functions may be found in Section 5, Problems 1–30, page 38.

2 Differentiation of products and quotients of two functions

(i) Differentiation of a product

The function $y = 3x^2 \sin x$ is a product of two terms in x, i.e. $3x^2$ and $\sin x$.

Let $u = 3x^2$ and $v = \sin x$.

Let x increase by a small increment δx, causing incremental changes in u, v and y of δu, δv and δy respectively.

Then
$$y = (3x^2)(\sin x)$$
$$= (u)(v)$$
$$y + \delta y = (u + \delta u)(v + \delta v)$$
$$= uv + v\delta u + u\delta v + \delta u\delta v$$
$$(y + \delta y) - (y) = uv + v\delta u + u\delta v + \delta u\delta v - uv$$
$$\delta y = v\delta u + u\delta v + \delta u\delta v$$

Dividing both sides by δx gives:

$$\frac{\delta y}{\delta x} = v\frac{\delta u}{\delta x} + u\frac{\delta v}{\delta x} + \frac{\delta u}{\delta x}\delta v$$

As $\delta x \to 0$ then $\delta u \to 0$, $\delta v \to 0$ and $\delta y \to 0$

However, the fact that δu and δx, for example, both approach zero does not mean that $\frac{\delta u}{\delta x}$ will approach zero.

Ratios of small quantities, such as $\frac{\delta u}{\delta x}$, $\frac{\delta v}{\delta x}$ or $\frac{\delta y}{\delta x}$ can be significant.

Consider two lines AB and AC meeting at A and whose intersecting angle (i.e. ∠ BAC) is any value.

If AB = δy = 1 unit, say, and AC = δx = 2 units, then the ratio

$$\frac{\delta y}{\delta x} = \frac{1}{2}.$$

This ratio of $\frac{1}{2}$ is still the same whether the unit of δy and δx is in, say, kilometres or millimetres. No matter how small δy or δx is made, the ratio is still $\frac{1}{2}$. Thus when $\delta y \to 0$ and when $\delta x \to 0$, the ratio $\frac{\delta y}{\delta x}$ is still a significant value.

As $\delta x \to 0$, $\frac{\delta u}{\delta x} \to \frac{du}{dx}$, $\frac{\delta v}{\delta x} \to \frac{dv}{dx}$, $\frac{\delta y}{\delta x} \to \frac{dy}{dx}$ and $\delta v \to 0$

Hence $\frac{dy}{dx} = v\frac{du}{dx} + u\frac{dv}{dx}$

This is known as the **product rule.**

Summary

When $y = uv$ and u and v are functions of x, then

$$\frac{dy}{dx} = v\frac{du}{dx} + u\frac{dv}{dx}$$

Using functional notation: When $F(x) = f(x)\,g(x)$ then:

$$F'(x) = f(x)\,g'(x) + g(x)\,f'(x)$$

Applying the product rule to $y = 3x^2 \sin x$:

let $u = 3x^2$ and $v = \sin x$

Then $\frac{dy}{dx} = (\sin x)\frac{d}{dx}(3x^2) + (3x^2)\frac{d}{dx}(\sin x)$

$= (\sin x)(6x) + (3x^2)(\cos x)$

$= 6x \sin x + 3x^2 \cos x$

i.e. $\frac{dy}{dx} = 3x\,(2 \sin x + x \cos x)$

From the above it should be noted that the differential coefficient of a product cannot be obtained merely by differentiating each term and multiplying the two answers together. The above formula must be used whenever differentiating products.

(ii) Differentiation of a quotient

The function $y = \frac{3 \cos x}{5x^3}$ is a quotient of two terms in x, i.e. $3 \cos x$ and $5x^3$.

Let $u = 3 \cos x$ and $v = 5x^3$.

Let x increase by a small increment δx causing incremental changes in u, v and y of δu, δv and δy respectively.

$$\text{Then} \quad y = \frac{3 \cos x}{5x^3}$$

$$= \frac{u}{v}$$

$$y + \delta y = \frac{u + \delta u}{v + \delta v}$$

$$(y + \delta y) - (y) = \frac{u + \delta u}{v + \delta v} - \frac{u}{v}$$

$$= \frac{uv + v\delta u - uv - u\delta v}{v(v + \delta v)}$$

$$\text{i.e.} \quad \delta y = \frac{v\delta u - u\delta v}{v^2 + v\delta v}$$

Dividing both sides by δx gives:

$$\frac{\delta y}{\delta x} = \frac{v\frac{\delta u}{\delta x} - u\frac{\delta v}{\delta x}}{v^2 + v\delta v}$$

As $\delta x \to 0$, $\frac{\delta u}{\delta x} \to \frac{du}{dx}$, $\frac{\delta v}{\delta x} \to \frac{dv}{dx}$, $\frac{\delta y}{\delta x} \to \frac{dy}{dx}$ and $\delta v \to 0$

$$\text{Hence} \quad \frac{dy}{dx} = \frac{v\frac{du}{dx} - u\frac{dv}{dx}}{v^2}$$

This is known as the **quotient rule.**

Summary

When $y = \frac{u}{v}$ and u and v are functions of x, then

$$\frac{dy}{dx} = \frac{v\frac{du}{dx} - u\frac{dv}{dx}}{v^2}$$

Using functional notation:

When $F(x) = \dfrac{f(x)}{g(x)}$, then $F'(x) = \dfrac{g(x) f'(x) - f(x) g'(x)}{[g(x)]^2}$

Applying the quotient rule to $y = \dfrac{3 \cos x}{5x^3}$:

Let $u = 3 \cos x$ and $v = 5x^3$

$$\text{Then } \frac{dy}{dx} = \frac{(5x^3)\frac{d}{dx}(3 \cos x) - (3 \cos x)\frac{d}{dx}(5x^3)}{(5x^3)^2}$$

$$= \frac{(5x^3)(-3 \sin x) - (3 \cos x)(15x^2)}{25x^6}$$

$$= \frac{-15x^2 (x \sin x + 3 \cos x)}{25x^6}$$

$$\text{i.e.} \quad \frac{dy}{dx} = \frac{-3(x \sin x + 3 \cos x)}{5x^4}$$

From above it should be noted that the differential coefficient of a quotient cannot be obtained by merely differentiating each term and dividing the numerator by the denominator. The above formula must be used when differentiating quotients.

The first step when differentiating a product such as $y = uv$ or a quotient such as $y = \dfrac{u}{v}$ is to decide clearly which is the u part and which is the v part. When this has been decided differentiation involves substitution into the appropriate formula.

Worked Problems on differentiating products and quotients

Problem 1. Find the differential coefficient of $5x^2 \cos x$.

Let $y = 5x^2 \cos x$

Also, let $u = 5x^2$ and $v = \cos x$

Then $\dfrac{du}{dx} = 10x$ and $\dfrac{dv}{dx} = -\sin x$

$$\text{Then } \frac{dy}{dx} = v\frac{du}{dx} + u\frac{dv}{dx}$$

$$= (\cos x)(10x) + (5x^2)(-\sin x)$$

$$= 10x \cos x - 5x^2 \sin x$$

$$\text{i.e.} \quad \frac{dy}{dx} = 5x(2 \cos x - x \sin x)$$

Problem 2. Differentiate $3e^{2h} \sin h$ with respect to h.

Let $F(h) = 3e^{2h} \sin h$
Let $f(h) = 3e^{2h}$ and $g(h) = \sin h$
then $f'(h) = 6e^{2h}$ and $g'(h) = \cos h$
Then $F'(h) = g(h)\, f'(h) + f(h)\, g'(h)$
$= (\sin h)\,(6e^{2h}) + (3e^{2h})\,(\cos h)$
$= 6e^{2h} \sin h + 3e^{2h} \cos h$
i.e. $F'(h) = 3e^{2h}\ [2 \sin h + \cos h]$

Problem 3. If $y = 7\sqrt{x} \ln 4x$ find $\dfrac{dy}{dx}$.

$y = 7x^{\frac{1}{2}} \ln 4x$

Let $u = 7x^{\frac{1}{2}}$ and $v = \ln 4x$

then $\dfrac{du}{dx} = \dfrac{7}{2} x^{-\frac{1}{2}}$ and $\dfrac{dv}{dx} = \dfrac{1}{x}$

$$\text{Then } \frac{dy}{dx} = v\frac{du}{dx} + u\frac{dv}{dx}$$

$$= (\ln 4x)\left[\frac{7}{2} x^{-\frac{1}{2}}\right] + [7x^{\frac{1}{2}}]\left[\frac{1}{x}\right]$$

$$= \frac{7}{2\sqrt{x}} \ln 4x + \frac{7}{\sqrt{x}}$$

$$\text{i.e. } \frac{dy}{dx} = \frac{7}{2\sqrt{x}} (\ln 4x + 2)$$

Problem 4. Find the differential coefficients of: (a) $\tan x$; (b) $\cot x$; (c) $\sec x$; (d) $\operatorname{cosec} x$.

(a) Let $y = \tan x = \dfrac{\sin x}{\cos x}$

Differentiation of $\tan x$ is treated as a quotient with $u = \sin x$ and $v = \cos x$.

Then $\dfrac{du}{dx} = \cos x$ and $\dfrac{dv}{dx} = -\sin x$

$$\frac{dy}{dx} = \frac{v\dfrac{du}{dx} - u\dfrac{dv}{dx}}{v^2}$$

$$= \frac{(\cos x)\,(\cos x) - (\sin x)\,(-\sin x)}{(\cos x)^2}$$

$$= \frac{(\cos^2 x + \sin^2 x)}{(\cos x)^2}$$

$$= \frac{1}{\cos^2 x} \text{ (since } \cos^2 x + \sin^2 x = 1)$$

i.e. $\frac{dy}{dx} = \sec^2 x$

Hence, when $y = \tan x$, $\frac{dy}{dx} = \sec^2 x$

or, when $f(x) = \tan x$, $f'(x) = \sec^2 x$

(b) Let $y = \cot x = \frac{\cos x}{\sin x}$

Differentiation of $\cot x$ is treated as a quotient with $u = \cos x$ and $v = \sin x$.

Then $\frac{du}{dx} = -\sin x$ and $\frac{dv}{dx} = \cos x$

$$\frac{dy}{dx} = \frac{v\frac{du}{dx} - u\frac{dv}{dx}}{v^2}$$

$$= \frac{(\sin x)(-\sin x) - (\cos x)(\cos x)}{(\sin x)^2}$$

$$= \frac{-(\sin^2 x + \cos^2 x)}{\sin^2 x}$$

$$= \frac{-1}{\sin^2 x}$$

i.e. $\frac{dy}{dx} = -\operatorname{cosec}^2 x$

Hence when $y = \cot x$, $\frac{dy}{dx} = -\operatorname{cosec}^2 x$

or, when $f(x) = \cot x$, $f'(x) = -\operatorname{cosec}^2 x$

(c) Let $y = \sec x = \frac{1}{\cos x}$

Differentiation of $\sec x$ is treated as a quotient with $u = 1$ and $v = \cos x$.

Then $\frac{du}{dx} = 0$ and $\frac{dv}{dx} = -\sin x$

$$\frac{dy}{dx} = \frac{v\frac{du}{dx} - u\frac{dv}{dx}}{v^2}$$

$$= \frac{(\cos x)(0) - (1)(-\sin x)}{(\cos x)^2}$$

$$= \frac{\sin x}{\cos^2 x}$$

$$= \left[\frac{1}{\cos x}\right]\left[\frac{\sin x}{\cos x}\right]$$

i.e. $\dfrac{dy}{dx} = \sec x \tan x$

Hence when $y = \sec x$, $\dfrac{dy}{dx} = \sec x \tan x$

or, when $f(x) = \sec x$, $f'(x) = \sec x \tan x$

(d) Let $y = \operatorname{cosec} x = \dfrac{1}{\sin x}$.

Differentiation of cosec x is treated as a quotient with $u = 1$ and $v = \sin x$.

Then $\dfrac{du}{dx} = 0$ and $\dfrac{dv}{dx} = \cos x$

$$\frac{dy}{dx} = \frac{v\dfrac{du}{dx} - u\dfrac{dv}{dx}}{v^2}$$

$$= \frac{(\sin x)(0) - (1)(\cos x)}{(\sin x)^2}$$

$$= \frac{-\cos x}{\sin^2 x}$$

$$= -\left[\frac{1}{\sin x}\right]\left[\frac{\cos x}{\sin x}\right]$$

i.e. $\dfrac{dy}{dx} = -\operatorname{cosec} x \cot x$

Hence when $y = \operatorname{cosec} x$, $\dfrac{dy}{dx} = -\operatorname{cosec} x \cot x$

or, when $f(x) = \operatorname{cosec} x$, $f'(x) = -\operatorname{cosec} x \cot x$

The differential coefficients of the six trigonometrical ratios may thus be summarised as below:

	y or $f(x)$	$\dfrac{dy}{dx}$ or $f'(x)$
1.	$\sin x$	$\cos x$
2.	$\cos x$	$-\sin x$
3.	$\tan x$	$\sec^2 x$
4.	$\sec x$	$\sec x \tan x$
5.	$\operatorname{cosec} x$	$-\operatorname{cosec} x \cot x$
6.	$\cot x$	$-\operatorname{cosec}^2 x$

Problem 5. If $f(t) = \dfrac{4e^{7t}}{\sqrt[3]{t^2}}$ find $f'(t)$

$$f(t) = \frac{4e^{7t}}{t^{\frac{2}{3}}}$$

Let $\quad g(t) = 4e^{7t}$ and $h(t) = t^{\frac{2}{3}}$

then $g'(t) = 28e^{7t}$ and $h'(t) = \frac{2}{3}t^{-\frac{1}{3}}$

$$f'(t) = \frac{h(t)\,g'(t) - g(t)\,h'(t)}{[h(t)]^2} = \frac{(t^{\frac{2}{3}})(28e^{7t}) - (4e^{7t})(\frac{2}{3}t^{-\frac{1}{3}})}{(t^{\frac{2}{3}})^2}$$

$$= \frac{28t^{\frac{2}{3}}e^{7t} - \frac{8}{3}t^{-\frac{1}{3}}e^{7t}}{t^{\frac{4}{3}}} = \frac{28t^{\frac{2}{3}}e^{7t}}{t^{\frac{4}{3}}} - \frac{8t^{-\frac{1}{3}}e^{7t}}{3t^{\frac{4}{3}}}$$

$$= 28t^{-\frac{2}{3}}e^{7t} - \frac{8}{3}t^{-\frac{5}{3}}e^{7t}$$

$$= \frac{4}{3}e^{7t}\,t^{-\frac{5}{3}}(21t - 2)$$

i.e. $\quad$ **$f'(t) = \dfrac{4e^{7t}}{3\sqrt[3]{t^5}}(21t - 2)$**

(Note that initially, $f(t) = \dfrac{4e^{7t}}{t^{\frac{2}{3}}}$ could have been treated as a product $f(t) = 4e^{7t}\,t^{-\frac{2}{3}}$)

Problem 6. Find the coordinates of the points on the curve $y = \dfrac{\frac{1}{3}(5 - 6x)}{3x^2 + 2}$ where the gradient is zero.

$$y = \frac{\frac{1}{3}(5 - 6x)}{3x^2 + 2}$$

Let $\quad u = \frac{1}{3}(5 - 6x)$ and $v = 3x^2 + 2$

then $\dfrac{du}{dx} = -2$ and $\dfrac{dv}{dx} = 6x$

$$\frac{dy}{dx} = \frac{v\dfrac{du}{dx} - u\dfrac{dv}{dx}}{v^2} = \frac{(3x^2 + 2)(-2) - \frac{1}{3}(5 - 6x)(6x)}{(3x^2 + 2)^2}$$

$$= \frac{-6x^2 - 4 - 10x + 12x^2}{(3x^2 + 2)^2} = \frac{6x^2 - 10x - 4}{(3x^2 + 2)^2}$$

When the gradient is zero, $\dfrac{dy}{dx} = 0$

Hence $\quad 6x^2 - 10x - 4 = 0$

$$2(3x^2 - 5x - 2) = 0$$

$$2(3x + 1)(x - 2) = 0$$

i.e. $x = -\frac{1}{3}$ or $x = 2$

Substituting in the original equation for y:

When $x = -\frac{1}{3}$, $y = \dfrac{\frac{1}{3}[5 - 6(-\frac{1}{3})]}{3(-\frac{1}{3})^2 + 2} = \dfrac{\frac{7}{3}}{\frac{7}{3}} = 1$

When $x = 2$, $y = \dfrac{\frac{1}{3}[5 - 6(2)]}{3(2)^2 + 2} = \dfrac{-\frac{7}{3}}{14} = -\frac{1}{6}$

Hence the coordinates of the points on the curve $y = \dfrac{\frac{1}{3}(5 - 6x)}{3x^2 + 2}$ where the gradient is zero are $(-\frac{1}{3}, 1)$ and $(2, -\frac{1}{6})$.

Problem 7. Differentiate $\dfrac{\sqrt{x}\sin x}{2e^{4x}}$ with respect to x.

The function $\dfrac{\sqrt{x}\sin x}{2e^{4x}}$ is a quotient, although the numerator (i.e. $\sqrt{x}\sin x$) is a product.

Let $\quad y = \dfrac{x^{\frac{1}{2}}\sin x}{2e^{4x}}$

Let $\quad u = x^{\frac{1}{2}}\sin x$ and $v = 2e^{4x}$

then $\quad \dfrac{du}{dx} = (x^{\frac{1}{2}})(\cos x) + (\sin x)(\frac{1}{2}x^{-\frac{1}{2}})$

and $\quad \dfrac{dv}{dx} = 8e^{4x}$

$$\frac{dy}{dx} = \frac{v\dfrac{du}{dx} - u\dfrac{dv}{dx}}{v^2}$$

$$= \frac{(2e^{4x})(x^{\frac{1}{2}}\cos x + \frac{1}{2}x^{-\frac{1}{2}}\sin x) - (x^{\frac{1}{2}}\sin x)(8e^{4x})}{(2e^{4x})^2}$$

Dividing throughout by $2e^{4x}$ gives:

$$\frac{dy}{dx} = \frac{x^{\frac{1}{2}}\cos x + \frac{1}{2}x^{-\frac{1}{2}}\sin x - 4x^{\frac{1}{2}}\sin x}{2e^{4x}}$$

Hence $\quad \dfrac{dy}{dx} = \dfrac{\sqrt{x}\cos x + \sin x\left(\dfrac{1}{2\sqrt{x}} - 4\sqrt{x}\right)}{2e^{4x}}$

or $\quad \dfrac{dy}{dx} = \dfrac{\sqrt{x}\cos x + \left(\dfrac{1 - 8x}{2\sqrt{x}}\right)\sin x}{2e^{4x}}$

Further problems on differentiating products and quotients may be found in Section 5, Problems 31–64, page 40.

3 Differentiation by substitution

The function $y = (4x - 3)^7$ can be differentiated by firstly multiplying $(4x - 3)$

by itself seven times, and then differentiating each term produced in turn. This would be a long process. In this type of function a substitution is made.

Let $u = 4x - 3$, then instead of $y = (4x - 3)^7$ we have $y = u^7$.

An important rule that is used when differentiating by substitution is:

$$\frac{dy}{dx} = \frac{dy}{du} \cdot \frac{du}{dx}$$

This is often known as the **chain rule.**

From above, $y = (4x - 3)^7$
If $u = 4x - 3$ then $y = u^7$

Thus $\frac{dy}{du} = 7u^6$ and $\frac{du}{dx} = 4$

Hence since $\frac{dy}{dx} = \frac{dy}{du} \cdot \frac{du}{dx}$

$$\frac{dy}{dx} = (7u^6)(4) = 28u^6$$

Rewriting $u = 4x - 3$, $\frac{dy}{dx} = 28(4x - 3)^6$

Since y is a function of u, and u is a function of x, then y is a 'function of a function' of x. The method of obtaining differential coefficients by making substitutions is often called the 'function of a function process'.

Worked problems on differentiation by substitution

Problem 1. Differentiate sin $(6x + 1)$.

Let $y = \sin(6x + 1)$
and $u = 6x + 1$

Then $y = \sin u$, giving $\frac{dy}{du} = \cos u$

and $\frac{du}{dx} = 6$

Using the 'differentiation by substitution' formula: $\frac{dy}{dx} = \frac{dy}{du} \cdot \frac{du}{dx}$ gives

$\frac{dy}{dx} = (\cos u)(6) = 6 \cos u$

Rewriting $u = 6x + 1$ gives:

$$\frac{dy}{dx} = 6\cos(6x + 1)$$

Note that this result could have been obtained by firstly differentiating the trigonometric function (i.e. differentiating sin $f(x)$) giving cos $f(x)$ and then multiplying by the differential coefficient of $f(x)$, i.e. 6.

Problem 2. Find the differential coefficient of $(3t^4 - 2t)^5$.

Let $\quad y = (3t^4 - 2t)^5$

and $\quad u = 3t^4 - 2t$

Then $\quad y = u^5$, giving $\dfrac{dy}{du} = 5u^4$

and $\quad \dfrac{du}{dt} = 12t^3 - 2$

Using the 'chain rule': $\dfrac{dy}{dt} = \dfrac{dy}{du} \cdot \dfrac{du}{dt}$ gives $\dfrac{dy}{dt} = (5u^4)(12t^3 - 2)$

Rewriting $u = 3t^4 - 2t$ gives:

$$\frac{dy}{dt} = 5(3t^4 - 2t)^4 (12t^3 - 2)$$

Note that this result could have been obtained by firstly differentiating the bracket, giving $5[f(x)]^4$ and then multiplying this result by the differential coefficient of $f(x)$ (i.e. $(12t^3 - 2)$).

Problem 3. If $y = 5 \operatorname{cosec}(3\sqrt{x} + 2x)$ find $\dfrac{dy}{dx}$.

$$y = 5 \operatorname{cosec}(3\sqrt{x} + 2x)$$

Let $\quad u = (3\sqrt{x} + 2x)$ then $\dfrac{du}{dx} = \dfrac{3}{2\sqrt{x}} + 2$

Thus $\quad y = 5 \operatorname{cosec} u$ and $\dfrac{dy}{du} = -5 \operatorname{cosec} u \cot u$

Now $\quad \dfrac{dy}{dx} = \dfrac{dy}{du} \cdot \dfrac{du}{dx} = (-5 \operatorname{cosec} u \cot u)\left(\dfrac{3}{2\sqrt{x}} + 2\right)$

Rewriting $u = 3\sqrt{x} + 2x$ gives:

$$\frac{dy}{dx} = -5\left(\frac{3}{2\sqrt{x}} + 2\right) \operatorname{cosec}(3\sqrt{x} + 2x) \cot(3\sqrt{x} + 2x)$$

In a similar way to Problem 1, this result could have been obtained by firstly differentiating $5 \operatorname{cosec} f(x)$ giving $-5 \operatorname{cosec} f(x) \cot f(x)$ and then multiplying this result by the differential coefficient of $f(x)$.

Problem 4. If $p = 2 \tan^5 v$ find $\dfrac{dp}{dv}$.

$$p = 2 \tan^5 v$$

Let $\quad u = \tan v$ then $\dfrac{du}{dv} = \sec^2 v$

Then $\quad p = 2u^5$ and $\dfrac{dp}{du} = 10u^4$

Now $\quad \dfrac{dp}{dv} = \dfrac{dp}{du} \cdot \dfrac{du}{dv} = (10u^4)(\sec^2 v)$

Rewriting $u = \tan v$ gives:

$$\frac{dp}{dv} = 10\,(\tan v)^4 \sec^2 v$$

$$\frac{dp}{dv} = 10 \tan^4 v \sec^2 v$$

In a similar way to Problem 2, this result could have been obtained by firstly differentiating the bracket (i.e. differentiating $2[f(v)]^5$) giving $10[f(v)]^4$ and then multiplying this result by the differential coefficient of $f(v)$.

Problem 5. Write down the differential coefficients of the following: (a) $\sqrt{(4x^2 + x - 3)}$; (b) $2 \sec^3 t$; (c) $4 \cot (5g^2 + 2)$; (d) $\sqrt{(4x^3 + 2)^3} \cos (3x^2 + 2)$.

(a) $f(x) = \sqrt{(4x^2 + x - 3)} = (4x^2 + x - 3)^{\frac{1}{2}}$

$$f'(x) = \tfrac{1}{2}\,(4x^2 + x - 3)^{-\frac{1}{2}}\,(8x + 1)$$

$$= \frac{8x + 1}{2\sqrt{(4x^2 + x - 3)}}$$

(b) $f(t) = 2 \sec^3 t = 2\,(\sec t)^3$

$$f'(t) = 6\,(\sec t)^2\,(\sec t \tan t)$$

$$= 6 \sec^3 t \tan t$$

(c) $f(g) = 4 \cot (5g^2 + 2)$

$$f'(g) = 4\,[-\operatorname{cosec}^2 (5g^2 + 2)]\,(10g)$$

$$= -40g \operatorname{cosec}^2 (5g^2 + 2)$$

(d) $f(x) = \sqrt{(4x^3 + 2)^3} \cos (3x^2 + 2)$

$$= (4x^3 + 2)^{\frac{3}{2}} \cos (3x^2 + 2) \quad \text{(i.e. a product)}$$

$$f'(x) = [\cos (3x^2 + 2)]\,[\tfrac{3}{2}\,(4x^3 + 2)^{\frac{1}{2}}\,(12x^2)] + [(4x^3 + 2)^{\frac{3}{2}}]\,[(-\sin (3x^2 + 2))(6x)]$$

$$= 6x\sqrt{(4x^3 + 2)}\,[3x \cos (3x^2 + 2) - (4x^3 + 2) \sin (3x^2 + 2)]$$

Further problems on differentiation by substitution may be found in Section 5, Problems 65–128, page 42.

4 Successive differentiation

When a function, say, $y = f(x)$, is differentiated, the differential coefficient is written as $f'(x)$ or $\frac{dy}{dx}$.

If the expression is differentiated again, the second differential coefficient or the second derivative is obtained. This is written as $f''(x)$ (pronounced 'f double-dash x') or $\frac{d^2y}{dx^2}$ (pronounced 'dee two y by dee x squared'). Similarly, if differentiated again the third differential coefficient or third derivative is obtained, and is written as $f'''(x)$ or $\frac{d^3y}{dx^3}$, and so on.

Worked problems on successive differentiation

Problem 1. If $f(x) = 3x^4 + 2x^3 + x - 1$, find $f'(x)$ and $f''(x)$.

$$f(x) = 3x^4 + 2x^3 + x - 1$$
$$f'(x) = (3)(4)x^3 + (2)(3)x^2 + 1 - 0$$
$$= 12x^3 + 6x^2 + 1$$
$$f''(x) = (12)(3)x^2 + (6)(2)x + 0$$
$$= 36x^2 + 12x$$

Problem 2. If $y = \frac{4}{3}x^3 - \frac{2}{x^2} + \frac{1}{3x} - \sqrt{x}$ find $\frac{d^2y}{dx^2}$ and $\frac{d^3y}{dx^3}$.

$$y = \frac{4}{3}x^3 - \frac{2}{x^2} + \frac{1}{3x} - \sqrt{x}$$
$$= \frac{4}{3}x^3 - 2x^{-2} + \frac{1}{3}x^{-1} - x^{\frac{1}{2}}$$

$$\frac{dy}{dx} = [\tfrac{4}{3}](3)\,x^2 - (2)(-2)x^{-3} + \tfrac{1}{3}(-1)x^{-2} - [\tfrac{1}{2}]\,x^{-\frac{1}{2}}$$
$$= 4x^2 + 4x^{-3} - \tfrac{1}{3}x^{-2} - \tfrac{1}{2}x^{-\frac{1}{2}}$$

$$\frac{d^2y}{dx^2} = (4)(2)x + (4)(-3)x^{-4} - [\tfrac{1}{3}](-2)x^{-3} - [\tfrac{1}{2}]\,[-\tfrac{1}{2}]\,x^{-\frac{3}{2}}$$
$$= 8x - 12x^{-4} + \tfrac{2}{3}x^{-3} + \tfrac{1}{4}x^{-\frac{3}{2}}$$

$$\text{i.e. } \frac{d^2y}{dx^2} = 8x - \frac{12}{x^4} + \frac{2}{3x^3} + \frac{1}{4\sqrt{x^3}}$$

$$\frac{d^3y}{dx^3} = 8 - (12)(-4)\,x^{-5} + [\tfrac{2}{3}](-3)x^{-4} + [\tfrac{1}{4}]\,[-\tfrac{3}{2}]\,x^{-\frac{5}{2}}$$
$$= 8 + 48x^{-5} - 2x^{-4} - \tfrac{3}{8}x^{-\frac{5}{2}}$$

$$\text{i.e. } \frac{d^3y}{dx^3} = 8 + \frac{48}{x^5} - \frac{2}{x^4} - \frac{3}{8\sqrt{x^5}}$$

Problem 3. Evaluate $f'(t)$ and $f''(t)$, correct to 3 decimal places when $t = \frac{1}{2}$ given $f(t) = 3 \ln \cos 2t$.

$$f(t) = 3 \ln \cos 2t$$

$$f'(t) = 3\left(\frac{1}{\cos 2t}\right)(-2 \sin 2t)$$

$$= -6 \tan 2t$$

When $t = \frac{1}{2}$, $f'(t) = -6 \tan 1 = -6\,(1.5574) = -9.344$

$f''(t) = -6\,(\sec^2 2t)\;2$

$= -12 \sec^2 2t$

When $t = \frac{1}{2}$, $f''(t) = -12\,(3.4255) = -41.106$

Problem 4. If $y = Ae^{2x} + Be^{-3x}$ prove that $\frac{d^2y}{dx^2} + \frac{dy}{dx} - 6y = 0$.

$y = Ae^{2x} + Be^{-3x}$

$\frac{dy}{dx} = 2Ae^{2x} - 3Be^{-3x}$

$\frac{d^2y}{dx^2} = 4Ae^{2x} + 9Be^{-3x}$

$6y = 6(Ae^{2x} + Be^{-3x}) = 6Ae^{2x} + 6Be^{-3x}$

Substituting into $\frac{d^2y}{dx^2} + \frac{dy}{dx} - 6y$ gives:

$(4Ae^{2x} + 9Be^{-3x}) + (2Ae^{2x} - 3Be^{-3x}) - (6Ae^{2x} + 6Be^{-3x})$

$= 4Ae^{2x} + 9Be^{-3x} + 2Ae^{2x} - 3Be^{-3x} - 6Ae^{2x} - 6Be^{-3x} = 0$

Thus $\frac{d^2y}{dx^2} + \frac{dy}{dx} - 6y = 0$

(Note that an equation of the form $\frac{d^2y}{dx^2} + \frac{dy}{dx} - 6y = 0$ is known as a 'differential equation' and such equations are discussed in Chapter 7.)

Further problems on successive differentiation may be found in the following section (5), Problems 129–151, page 44.

5 Further problems

Differentiation of common functions

Find the differential coefficients with respect to x of the functions in Problems 1–6.

1. (a) x^4 (b) x^6 (c) x^9 (d) $x^{3.2}$ (e) $x^{4.7}$
(a) $[4x^3]$ (b) $[6x^5]$ (c) $[9x^8]$ (d) $[3.2x^{2.2}]$ (e) $[4.7x^{3.7}]$

2. (a) $3x^3$ (b) $4x^7$ (c) $2x^{10}$ (d) $4.6x^{1.5}$ (e) $6x^{5.4}$
(a) $[9x^2]$ (b) $[28x^6]$ (c) $[20x^9]$ (d) $[6.9x^{0.5}]$
(e) $[32.4x^{4.4}]$

3. (a) x^{-2} (b) x^{-3} (c) x^{-5} (d) $\frac{1}{x}$ (e) $-\frac{1}{x^3}$ (f) $\frac{1}{x^{10}}$

(a) $[-2x^{-3}]$ (b) $[-3x^{-4}]$ (c) $[-5x^{-6}]$ (d) $\left[-\frac{1}{x^2}\right]$
(e) $\left[\frac{3}{x^4}\right]$ (f) $\left[-\frac{10}{x^{11}}\right]$

4. (a) $4x^{-1}$ (b) $-5x^{-4}$ (c) $3x^{-7}$ (d) $-\dfrac{6}{x^2}$ (e) $\dfrac{4}{3x^5}$ (f) $\dfrac{2}{5x^{1.4}}$

(a) $[-4x^{-2}]$ (b) $[20x^{-5}]$ (c) $[-21x^{-8}]$ (d) $\left[\dfrac{12}{x^3}\right]$

(e) $\left[-\dfrac{20}{3x^6}\right]$ (f) $\left[\dfrac{-2.8}{5x^{2.4}}\right]$

5. (a) $x^{\frac{7}{2}}$ (b) $x^{\frac{3}{4}}$ (c) $x^{-\frac{3}{2}}$ (d) $\dfrac{1}{x^{\frac{1}{2}}}$ (e) $-\dfrac{1}{x^{\frac{4}{3}}}$ (f) $\dfrac{2}{3x^{\frac{7}{4}}}$

(a) $[\frac{7}{2}x^{\frac{5}{2}}]$ (b) $[\frac{3}{4}x^{-\frac{1}{4}}]$ (c) $[-\frac{3}{2}x^{-\frac{5}{2}}]$ (d) $\left[\dfrac{-1}{2x^{\frac{3}{2}}}\right]$ (e) $\left[\dfrac{4}{3x^{\frac{7}{3}}}\right]$

(f) $\left[\dfrac{-7}{6x^{\frac{11}{4}}}\right]$

6. (a) $\dfrac{\sqrt{x}}{2}$ (b) $\sqrt{x^3}$ (c) $\sqrt[3]{x^2}$ (d) $4\sqrt{x^5}$ (e) $\dfrac{3}{5\sqrt{x^7}}$ (f) $\dfrac{-1}{2\sqrt[4]{x^9}}$

(a) $\left[\dfrac{1}{4\sqrt{x}}\right]$ (b) $\left[\dfrac{3}{2}\sqrt{x}\right]$ (c) $\left[\dfrac{2}{3\sqrt[3]{x}}\right]$ (d) $[10\sqrt{x^3}]$

(e) $\left[\dfrac{-21}{10\sqrt{x^9}}\right]$ (f) $\left[\dfrac{9}{8\sqrt[4]{x^{13}}}\right]$

Differentiate the functions in Problems 7–26 with respect to the variable:

7. (a) $4u^3$ (b) $\frac{3}{2}t^4$ (a) $[12u^2]$ (b) $[6t^3]$
8. (a) $5v^2$ (b) $1.4z^5$ (a) $[10v]$ (b) $[7z^4]$
9. (a) $\dfrac{4}{a}$ (b) $\dfrac{3}{2S^2}$ (a) $\left[-\dfrac{4}{a^2}\right]$ (b) $\left[-\dfrac{3}{S^3}\right]$
10. (a) $\dfrac{7}{4y^3}$ (b) $3m^{-4}$ (a) $\left[-\dfrac{21}{4y^4}\right]$ (b) $\left[-12m^{-5}\right]$
11. (a) $\sqrt{b}$ (b) $5\sqrt{c^3}$ (a) $\left[\dfrac{1}{2\sqrt{b}}\right]$ (b) $\left[\dfrac{15}{2}\sqrt{c}\right]$
12. (a) $\dfrac{1}{\sqrt{e}}$ (b) $g^{\frac{5}{3}}$ (a) $\left[-\dfrac{1}{2\sqrt{e^3}}\right]$ (b) $[\frac{5}{3}g^{\frac{2}{3}}]$
13. (a) $4\sqrt[3]{k^2}$ (b) $\dfrac{3}{5\sqrt[4]{x^5}}$ (a) $\left[\dfrac{8}{3\sqrt[3]{k}}\right]$ (b) $\left[\dfrac{-3}{4\sqrt[4]{x^9}}\right]$
14. $5x^2 - \dfrac{1}{\sqrt{x^7}}$ $\left[10x + \dfrac{7}{2\sqrt{x^9}}\right]$
15. $3\left(2u - u^{-\frac{1}{2}} + \dfrac{4}{5u}\right)$ $\left[3\left(2 + \dfrac{u^{-\frac{3}{2}}}{2} - \dfrac{4}{5u^2}\right)\right]$
16. $\dfrac{1}{x}\left(3x^3 - \dfrac{2}{x} + \dfrac{\sqrt{x}}{5} + 1\right)$ $\left[6x + \dfrac{4}{x^3} - \dfrac{1}{10\sqrt{x^3}} - \dfrac{1}{x^2}\right]$
17. $\dfrac{3x^2 - 2\sqrt{x} - 5\sqrt[4]{x^3}}{x^2}$ $\left[\dfrac{3}{\sqrt{x^5}} + \dfrac{25}{4\sqrt[4]{x^9}}\right]$

18. $(t + 1)^2$ $[2(t + 1)]$
19. $(3\theta - 1)^2$ $[6(3\theta - 1)]$
20. $(f - 1)^4$ $[4(f^3 - 3f^2 + 3f - 1)$ or $4(f - 1)^3]$
21. (a) $5 \sin \theta$ (b) $4 \cos x$ (a) $[5 \cos \theta]$ (b) $[-4 \sin x]$
22. (a) $3(\sin t + 2 \cos t)$ (b) $7 \sin x - 2 \cos x$
    (a) $[3(\cos t - 2 \sin t)]$ (b) $[7 \cos x + 2 \sin x]$
23. (a) e^{3x} (b) e^{-4y} (a) $[3e^{3x}]$ (b) $[-4e^{-4y}]$
24. (a) $6e^{2x}$ (b) $\dfrac{4}{e^{7t}}$ (a) $[12e^{2x}]$ (b) $\left[\dfrac{-28}{e^{7t}}\right]$
25. (a) $3(e^{8y} - e^{3y})$ (b) $-2(3e^{9x} - 4e^{-2x})$
    (a) $[3(8e^{8y} - 3e^{3y})]$ (b) $[-2(27e^{9x} + 8e^{-2x})]$
26. (a) $\ln 5b$ (b) $4 \ln 3g$ (a) $\left[\dfrac{1}{b}\right]$ (b) $\left[\dfrac{4}{g}\right]$
27. Find the gradient of the curve $y = 4x^3 - 3x^2 + 2x - 4$ at the points $(0, -4)$ and $(1, -1)$. [2, 8]
28. What are the coordinates of the point on the graph of $y = 5x^2 - 2x + 1$ where the gradient is zero. $[(\frac{1}{5}, \frac{4}{5})]$
29. Find the point on the curve $f(\theta) = 4\sqrt[3]{\theta^4} + 2$ where the gradient is $10\frac{2}{3}$. $[(8, 66)]$
30. If $f(x) = \dfrac{5x^2}{2} - 6x + 3$ find the coordinates at the point at which the gradient is: (a) 4; and (b) -6. (a) $[(2, 1)]$ (b) $[(0, 3)]$

Differentiation of products and quotients

Differentiate the products in Problems 31–45 with respect to the variable and express your answers in their simplest form:

31. $3x^3 \sin x$ $[3x^2(x \cos x + 3 \sin x)]$
32. $\sqrt{t^3} \cos t$ $[\sqrt{t}(\frac{3}{2} \cos t - t \sin t)]$
33. $(3x^2 - 4x + 2)(2x^3 + x - 1)$ $[(30x^4 - 32x^3 + 21x^2 - 14x + 6)]$
34. $2 \sin \theta \cos \theta$ $[2(\cos^2 \theta - \sin^2 \theta)]$
35. $5e^{2a} \sin a$ $[5e^{2a}(\cos a + 2 \sin a)]$
36. $e^{7y} \cos y$ $[e^{7y}(7 \cos y - \sin y)]$
37. $b^3 \ln 2b$ $[b^2(1 + 3 \ln 2b)]$
38. $3\sqrt{x}e^{4x}$ $\left[3e^{4x}\left(\dfrac{8x + 1}{2\sqrt{x}}\right)\right]$
39. $e^t \ln t$ $\left[e^t\left(\dfrac{1}{t} + \ln t\right)\right]$
40. $e^{2d}(4d^2 - 3d + 1)$ $[e^{2d}(8d^2 + 2d - 1)]$
41. $3\sqrt{f^5} \ln 5f$ $[3\sqrt{f^3}(1 + \frac{5}{2} \ln 5f)]$
42. $2 \sin g \ln g$ $\left[2\left(\dfrac{1}{g} \sin g + \ln g \cos g\right)\right]$
43. $6e^{5m} \sin m$ $[6e^{5m}(\cos m + 5 \sin m)]$
44. $\sqrt{x}(1 + \sin x)$ $\left[\dfrac{2x \cos x + \sin x + 1}{2\sqrt{x}}\right]$

45. $e^v \ln v \sin v$ $\left[e^v\left\{(\sin v + \cos v)\ln v + \frac{\sin v}{v}\right\}\right]$

Differentiate the quotients in Problems 46–62 with respect to the variable and express your answers in their simplest form:

46. $\frac{4x}{x^2-1}$ $\left[\frac{-4(x^2+1)}{(x^2-1)^2}\right]$

47. $\frac{2t-1}{3t^2+5t}$ $\left[\frac{5+6t-6t^2}{(3t^2+5t)^2}\right]$

48. $\frac{2x^2-6x+2}{3x^2+2x-1}$ $\left[\frac{2(11x^2-8x+1)}{(3x^2+2x-1)^2}\right]$

49. $\frac{3e^{2\theta}}{4\theta^2-3}$ $\left[\frac{6e^{2\theta}(4\theta^2-4\theta-3)}{(4\theta^2-3)^2}\right]$

50. $\frac{3u^4+2u^2-1}{4e^{5u}}$ $\left[\frac{-15u^4+12u^3-10u^2+4u+5}{4e^{5u}}\right]$

51. $\frac{4\sin c}{5c^2+2c}$ $\left[\frac{4(5c^2+2c)\cos c-4(10c+2)\sin c}{(5c^2+2c)^2}\right]$

52. $\frac{4\sqrt[3]{f^7}}{3\sin f}$ $\left[\frac{4(\sqrt[3]{f^4})(7\sin f-3f\cos f)}{9\sin^2 f}\right]$

53. $\frac{6\cos b}{b^3+4}$ $\left[\frac{-6\{(b^3+4)\sin b+3b^2\cos b\}}{(b^3+4)^2}\right]$

54. $\frac{\sqrt{k^3}}{\cos k}$ $\left[\frac{\sqrt{k}(\frac{3}{2}\cos k+k\sin k)}{\cos^2 k}\right]$

55. $\frac{4e^{6x}}{\sin x}$ $\left[\frac{4e^{6x}(6\sin x-\cos x)}{\sin^2 x}\right]$

56. $\frac{3\ln\frac{5}{2}n}{n^2+2n}$ $\left[\frac{3(n+2)-6(n+1)\ln\frac{5n}{2}}{(n^2+2n)^2}\right]$

57. $\frac{3\sqrt{x}+x}{\frac{7}{2}\ln 4x}$ $\left[\frac{\left(\frac{3}{2\sqrt{x}}+1\right)\ln 4x-\left(\frac{3}{\sqrt{x}}+1\right)}{\frac{7}{2}(\ln 4x)^2}\right]$

58. $\frac{\ln 6y}{6\sin y}$ $\left[\frac{\frac{1}{y}\sin y-\ln 6y\cos y}{6\sin^2 y}\right]$

59. $\frac{x^2\ln 4x}{3\sin x}$ $\left[\frac{x\ln 4x(2\sin x-x\cos x)+x\sin x}{3\sin^2 x}\right]$

60. $\frac{2\sqrt{t}}{\ln 3t\cos t}$ $\left[\frac{(\ln 3t\cos t+2t\ln 3t\sin t-2\cos t)}{\sqrt{t}(\ln 3t\cos t)^2}\right]$

61. $\frac{x^2\sec x}{e^{2x}}$ $\left[\frac{x\sec x}{e^{2x}}(x\tan x+2-2x)\right]$

62. $\frac{k}{e^k\operatorname{cosec} k}$ $\left[\frac{1+k(\cot k-1)}{e^k\operatorname{cosec} k}\right]$

63. Find the slope of the curve $y = xe^{-2x}$ at the point $\left(\frac{1}{2}, \frac{1}{2e}\right)$ [0]

64. Calculate the gradient of the curve $f(x) = \dfrac{3x^4 - 2\sqrt{x^3} + 2}{5x^2 + 1}$ at the points $(0, 2)$ and $(1, \frac{1}{2})$. $[0, \frac{2}{3}]$

Differentiation by substitution

Find the differential coefficients of the functions in Problems 65–128 with respect to the variable and express your answers in their simplest form.

65. $\sin 4x$ $[4 \cos 4x]$

66. $3 \tan 4x$ $[12 \sec^2 4x]$

67. $\cos 3t$ $[-3 \sin 3t]$

68. $5 \sec 2\theta$ $[10 \sec 2\theta \tan 2\theta]$

69. $4 \operatorname{cosec} 5\mu$ $[-20 \operatorname{cosec} 5\mu \cot 5\mu]$

70. $6 \cot 3\alpha$ $[-18 \operatorname{cosec}^2 3\alpha]$

71. $4 \cos (2x - 5)$ $[-8 \sin(2x - 5)]$

72. $\operatorname{cosec} (5t-1)$ $[-5 \operatorname{cosec} (5t - 1) \cot (5t - 1)]$

73. $(t^3 - 2t + 3)^4$ $[4(t^3 - 2t + 3)^3 (3t^2 - 2)]$

74. $\sqrt{(2v^3 - v)}$ $\left[\dfrac{6v^2 - 1}{2\sqrt{(2v^3 - v)}}\right]$

75. $\sin (3x - 2)$ $[3 \cos (3x - 2)]$

76. $3 \tan (5y - 1)$ $[15 \sec^2 (5y - 1)]$

77. $4 \cos (6x + 5)$ $[-24 \sin (6x + 5)]$

78. $(1 - 2u^2)^7$ $[-28u(1 - 2u^2)^6]$

79. $\dfrac{1}{2n^2 - 3n + 1}$ $\left[\dfrac{3 - 4n}{(2n^2 - 3n + 1)^2}\right]$

80. $\sin^2 t$ $[2 \sin t \cos t]$

81. $3 \cos^2 x$ $[-6 \cos x \sin x]$

82. $\dfrac{1}{(2g - 1)^6}$ $\left[\dfrac{-12}{(2g - 1)^7}\right]$

83. $3 \operatorname{cosec}^2 x$ $[-6 \operatorname{cosec}^2 x \cot x]$

84. $6 \cos^3 t$ $[-18 \cos^2 t \sin t]$

85. $\frac{3}{2} \cot (6x - 2)$ $[-9 \operatorname{cosec}^2 (6x - 2)]$

86. $\sqrt{(4x^3 + 2x^2 - 5x)}$ $\left[\dfrac{12x^2 + 4x - 5}{2\sqrt{(4x^3 + 2x^2 - 5x)}}\right]$

87. $2 \sin^4 b$ $[8 \sin^3 b \cos b]$

88. $\dfrac{3}{(x^2 + 6x - 1)^5}$ $\left[\dfrac{-30(x + 3)}{(x^2 + 6x - 1)^6}\right]$

89. $(x^2 - x + 1)^{12}$ $[12(x^2 - x + 1)^{11} (2x - 1)]$

90. $e^{4q + 3}$ $[4e^{4q + 3}]$

91. $5 e^{x-5}$ $[5 e^{x-5}]$

92. $\ln (3p - 1)$ $\left[\dfrac{3}{3p - 1}\right]$

93. $15 \ln \left(\dfrac{x}{3} + 5\right)$ $\left[\dfrac{15}{x + 15}\right]$

94. $3 \sec 5g$ $[15 \sec 5g \tan 5g]$

95. $4 \operatorname{cosec}(2k-1)$ $\quad [-8 \operatorname{cosec}(2k-1)\cot(2k-1)]$

96. $7\beta \tan 4\beta$ $\quad [7(4\beta \sec^2 4\beta + \tan 4\beta)]$

97. $\sqrt{x} \sec\frac{x}{3}$ $\quad \left[\frac{\sqrt{x}}{3}\sec\frac{x}{3}\tan\frac{x}{3} + \frac{1}{2\sqrt{x}}\sec\frac{x}{3}\right]$

98. $2\,e^{5l} \operatorname{cosec} 3l$ $\quad [2\,e^{5l} \operatorname{cosec} 3l(5 - 3\cot 3l)]$

99. $\ln 5v \cot v$ $\quad \left[\frac{\cot v}{v} - \operatorname{cosec}^2 v \ln 5v\right]$

100. $(3S^4 - 2S^2 + 1)\tan\frac{2S}{5}$

$\left[\frac{2}{5}(3S^4 - 2S^2 + 1)\sec^2\frac{2S}{5} + 4S(3S^2 - 1)\tan\frac{2S}{5}\right]$

101. $\frac{\sec 2t}{(t-1)}$ $\quad \left[\frac{\sec 2t}{(t-1)^2}\left\{2(t-1)\tan 2t - 1\right\}\right]$

102. $(x^2+1)\sin(2x^2-3)$ $\quad [2x\{2(x^2+1)\cos(2x^2-3) + \sin(2x^2-3)\}]$

103. $(2x-1)^9 \cos 4x$ $\quad [-2(2x-1)^8\{2(2x-1)\sin 4x - 9\cos 4x\}]$

104. $\frac{2t}{\tan 3t}$ $\quad \left[\frac{2(\tan 3t - 3t\sec^2 3t)}{\tan^2 3t}\right]$

105. $\frac{1}{\operatorname{cosec}(4v+1)}$ $\quad [4\cos(4v+1)]$

106. $\frac{\sin(6x-5)}{\sqrt{(x^2-1)}}$ $\quad \left[\frac{6(x^2-1)\cos(6x-5) - x\sin(6x-5)}{\sqrt{(x^2-1)^3}}\right]$

107. $\frac{2\sqrt{u}}{3\operatorname{cosec} 4u}$ $\quad \left[\frac{1 + 8u\cot 4u}{3\sqrt{u}\operatorname{cosec} 4u}\right]$

108. $\frac{3\cot x}{\ln 2x}$ $\quad \left[\frac{-(3\operatorname{cosec}^2 x \ln 2x + \frac{3}{x}\cot x)}{(\ln 2x)^2}\right]$

109. $\frac{5\tan 3b}{\sqrt{b}}$ $\quad \left[\frac{5}{2\sqrt{b^3}}(6b\sec^2 3b - \tan 3b)\right]$

110. $\frac{(3\theta^2-2)}{4\sec 2\theta}$ $\quad \left[\frac{3\theta - (3\theta^2-2)\tan 2\theta}{2\sec 2\theta}\right]$

111. $\frac{3\sec^2 a}{(2a^2+3a-1)^3}$ $\quad \left[\frac{3\sec^2 a[2(2a^2+3a-1)\tan a - 3(4a+3)]}{(2a^2+3a-1)^4}\right]$

112. $\frac{3e^{7x-1}}{(x-1)^9}$ $\quad \left[\frac{3e^{7x-1}(7x-16)}{(x-1)^{10}}\right]$

113. $\frac{\sqrt{(3c^2+4c-1)}}{2\ln 5c}$ $\quad \left[\frac{2c(3c+2)\ln 5c - 2(3c^2+4c-1)}{4c\sqrt{(3c^2+4c-1)}(\ln 5c)^2}\right]$

114. $3\cos(2x^2+1)$ $\quad [-12x\sin(2x^2+1)]$

115. $5\sec(5x^3-2x^2+2)$

$[5x(15x-4)\sec(5x^3-2x^2+2)\tan(5x^3-2x^2+2)]$

116. $\sin^2(2d-1)$ $\quad [4\sin(2d-1)\cos(2d-1)]$

117. $3\tan\sqrt{(4x-2)}$ $\quad \left[\frac{6\sec^2\sqrt{(4x-2)}}{\sqrt{(4x-2)}}\right]$

118. $3\cot^3(5t^2-6)$ $[-90t\cot^2(5t^2-6)\operatorname{cosec}^2(5t^2-6)]$

119. $\sin^4\sqrt{(3f^3-5)}$ $\left[\dfrac{18f^2\sin^3\sqrt{(3f^3-5)}\cos\sqrt{(3f^3-5)}}{\sqrt{(3f^3-5)}}\right]$

120. $e^{\sec g}$ $[\sec g\tan g\, e^{\sec g}]$

121. $3e^{\operatorname{cosec}(2x-1)}$ $[-6\operatorname{cosec}(2x-1)\cot(2x-1)\,e^{\operatorname{cosec}(2x-1)}]$

122. $3\sqrt[3]{\{\sin(4k-2)\}^5}$ $[20\sqrt[3]{\{\sin(4k-2)\}^2}\cos(4k-2)]$

123. $(3x^3+2x)^2\sin\sqrt{(x^2-1)}$

$\left[(3x^3+2x)\left\{\dfrac{x(3x^3+2x)}{\sqrt{(x^2-1)}}\cos\sqrt{(x^2-1)}+2(9x^2+2)\sin\sqrt{(x^2-1)}\right\}\right]$

124. $(x+3)^9\cos^4(x^2+2)$

$[(x+3)^8\cos^3(x^2+2)\{9\cos(x^2+2)-8x(x+3)\sin(x^2+2)\}]$

125. $\dfrac{3\cot^2 3m}{(2m-1)^5}$ $\left[\dfrac{-6\cot 3m}{(2m-1)^6}\left\{3(2m-1)\operatorname{cosec}^2 3m+5\cot 3m\right\}\right]$

126. $\dfrac{\sqrt{(3x^2-2)}}{\operatorname{cosec}^2(3x^2-2)}$ $\left[\dfrac{3x\{1+4(3x^2-2)\cot(3x^2-2)\}}{\sqrt{(3x^2-2)}\operatorname{cosec}^2(3x^2-2)}\right]$

127. $\ln\sqrt{(\operatorname{cosec} t)}$ $[-\frac{1}{2}\cot t]$

128. $\dfrac{3e^{3x^2+2x+1}}{\ln(\cos x)}$ $\left[\dfrac{3e^{3x^2+2x+1}}{[\ln(\cos x)]^2}\left\{(6x+2)\ln(\cos x)+\tan x\right\}\right]$

Successive differentiation

129. If $y=5x^3-6x^2+2x-6$ find $\dfrac{d^2y}{dx^2}$. $[30x-12]$

130. Find $f''(x)$ given $f(x)=\dfrac{5}{x}+\sqrt{x}-\dfrac{5}{\sqrt{x^5}}+8$. $\left[\dfrac{10}{x^3}-\dfrac{1}{4\sqrt{x^3}}-\dfrac{175}{4\sqrt{x^9}}\right]$

131. Given $f(\theta)=3\sin 4\theta-2\cos 3\theta$ find $f'(\theta)$, $f''(\theta)$ and $f'''(\theta)$.
$[f'(\theta)=6(2\cos 4\theta+\sin 3\theta); f''(\theta)=6(3\cos 3\theta-8\sin 4\theta);$
$f'''(\theta)=-6(32\cos 4\theta+9\sin 3\theta)]$

132. If $m=(6p+1)\left(\dfrac{1}{p}-3\right)$ find $\dfrac{d^2m}{dp^2}$ and $\dfrac{d^3m}{dp^3}$. $\left[\dfrac{2}{p^3};\dfrac{-6}{p^4}\right]$

In Problems 133–143 find the second differential coefficient with respect the variable.

133. $3\ln 5g$ $\left[-\dfrac{3}{g^2}\right]$

134. $(x-2)^5$ $[20(x-2)^3]$

135. $3\sin t-\cos 2t$ $[4\cos 2t-3\sin t]$

136. $3\tan 2y+4\cot 3y$ $[24(\sec^2 2y\tan 2y+3\operatorname{cosec}^2 3y\cot 3y)]$

137. $(3m^2-2)^6$ $[36(3m^2-2)^4(33m^2-2)]$

138. $\dfrac{1}{(2r-1)^7}$ $\left[\dfrac{224}{(2r-1)^9}\right]$

139. $3\cos^2\theta$ $[6(\sin^2\theta-\cos^2\theta)]$

140. $\frac{1}{2}\cot(3x-1)$ $[9\operatorname{cosec}^2(3x-1)\cot(3x-1)]$

141. $4\sin^5 n$ $[20\sin^3 n(4\cos^2 n-\sin^2 n)]$

142. $3x^2 \sin 2x$ $\quad [6(1-2x^2)\sin 2x + 24x\cos 2x]$

143. $\dfrac{\sin t}{2t^2}$ $\quad \left[\dfrac{1}{2t^4}\left\{(6-t^2)\sin t - 4t\cos t\right\}\right]$

144. $x = 3t^2 - 2\sqrt{t} + \dfrac{1}{t} - 6$. Evaluate $\dfrac{d^2x}{dt^2}$ when $t = 1$. $\quad [8\frac{1}{2}]$

145. Evaluate $f''(\theta)$ when $\theta = 0$ given $f(\theta) = 5\sec 2\theta$. $\quad [20]$

146. If $y = \cos\alpha - \sin\alpha$ evaluate α when $\dfrac{d^2y}{d\alpha^2}$ is zero. $\quad \left[\dfrac{\pi}{4}\right]$

147. If $y = Ae^x - Be^{-x}$ prove that: $\dfrac{e^x}{2}\left\{\dfrac{d^2y}{dx^2} + \dfrac{dy}{dx}\right\} - e^x y = B$.

148. Show that $\dfrac{d^2b}{dS^2} + 6\dfrac{db}{dS} + 25b = 0$ when $b = e^{-3S}\sin 4S$.

149. Show that $x = 2t\,e^{-2t}$ satisfies the equation: $\dfrac{d^2x}{dt^2} + 4\dfrac{dx}{dt} + 4x = 0$.

150. If $y = 3x^3 + 2x - 4$ prove that: $\dfrac{d^3y}{dx^3} + \dfrac{2}{9}\dfrac{d^2y}{dx^2} + x\dfrac{dy}{dx} - 3y = 30$.

151. Show that the differential equation $\dfrac{d^2y}{dx^2} - 8\dfrac{dy}{dx} + 41y = 0$ is satisfied when $y = 2e^{4x}\cos 5x$.

Chapter 3

Applications of differentiation

1 Velocity and acceleration

Let a car move a distance x metres in a time t seconds along a straight road. If the velocity v of the car is constant then

$$v = \frac{x}{t} \text{ m s}^{-1}$$

i.e. the gradient of the distance/time graph shown in Fig. 1 (a) is constant. If, however, the velocity of the car is not constant then the distance/time graph will not be a straight line. It may be as shown in Fig. 1 (b).

The average velocity over a small time δt and distance δx is given by the gradient of the chord CD, i.e. the average velocity over time $\delta t = \frac{\delta x}{\delta t}$.

As $\delta t \to 0$, the chord CD becomes a tangent, such that at point C the velocity v is given by:

$$v = \frac{\mathrm{d}x}{\mathrm{d}t}$$

Hence the velocity of the car at any instant t is given by the gradient of the distance/time graph. If an expression for the distance x is known in terms of time t then the velocity is obtained by differentiating the expression.

The acceleration a of the car is defined as the rate of change of velocity.

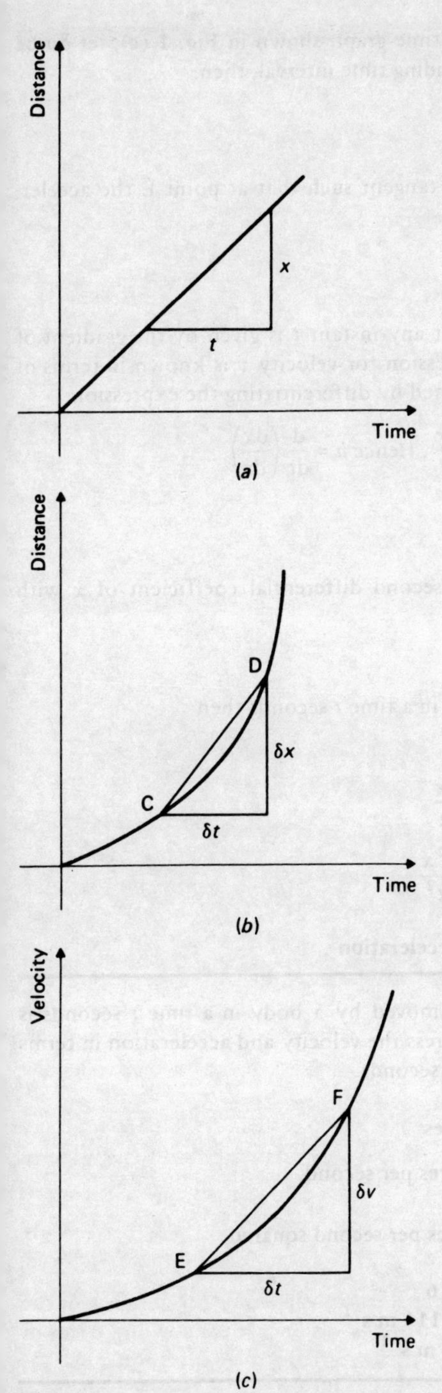
Distance
x
t
Time
(a)
Distance
D
δx
C
δt
Time
(b)
Velocity
F
δv
E
δt
Time
(c)

Fig. 1

With reference to the velocity/time graph shown in Fig. 1 (c), let δv be the change in v and δt the corresponding time interval, then:

$$\text{average acceleration } a = \frac{\delta v}{\delta t}$$

As $\delta t \to 0$, the chord EF becomes a tangent such that at point E the acceleration is given by:

$$a = \frac{\mathrm{d}v}{\mathrm{d}t}$$

Hence the acceleration of the car at any instant t is given by the gradient of the velocity/time graph. If an expression for velocity v is known in terms of time t then the acceleration is obtained by differentiating the expression.

$$\text{Acceleration, } a = \frac{\mathrm{d}v}{\mathrm{d}t}\text{; but } v = \frac{\mathrm{d}x}{\mathrm{d}t}\text{. Hence } a = \frac{\mathrm{d}}{\mathrm{d}t}\left(\frac{\mathrm{d}x}{\mathrm{d}t}\right)$$

$$\text{which is written as: } a = \frac{\mathrm{d}^2x}{\mathrm{d}t^2}$$

Thus acceleration is given by the second differential coefficient of x with respect to t.

Summary

If a body moves a distance x metres in a time t seconds then:

$$\text{distance } x = f(t)$$

$$\text{velocity } v = f'(t) \text{ or } \frac{\mathrm{d}x}{\mathrm{d}t}$$

$$\text{and acceleration } a = f''(t) \text{ or } \frac{\mathrm{d}^2x}{\mathrm{d}t^2}$$

Worked problems on velocity and acceleration

Problem 1. The distance x metres moved by a body in a time t seconds is given by $x = 2t^3 + 3t^2 - 6t + 2$. Express the velocity and acceleration in terms of t and find their values when $t = 4$ seconds.

Distance $x = 2t^3 + 3t^2 - 6t + 2$ metres

Velocity $v = \frac{\mathrm{d}x}{\mathrm{d}t} = 6t^2 + 6t - 6$ metres per second

Acceleration $a = \frac{\mathrm{d}^2x}{\mathrm{d}t^2} = 12t + 6$ metres per second squared

After 4 seconds, $v = 6(4)^2 + 6(4) - 6$
$= 96 + 24 - 6 =$ **114 m s^{-1}**
$a = 12(4) + 6 =$ **54 m s^{-2}**

Problem 2. If the distance s metres travelled by a car in time t seconds after the brakes are applied is given by $s = 15t - \frac{5}{3}t^2$: (a) what is the speed (in km h^{-1}) at the instant the brakes are applied, and (b) how far does the car travel before it stops?

(a) Distance $s = 15t - \frac{5}{3}t^2$

Velocity $v = \dfrac{ds}{dt} = 15 - \frac{10}{3}t$

At the instant the brakes are applied, $t = 0$.

Hence velocity = 15 m s^{-1}

$15 \text{ m s}^{-1} = \dfrac{15}{1\,000}(60 \times 60) \text{ km h}^{-1} = \mathbf{54 \text{ km h}^{-1}}$

(b) When the car finally stops, the velocity is zero,

i.e. $v = 15 - \frac{10}{3}t = 0$

i.e. $15 = \frac{10}{3}t$ or $t = 4.5$ seconds

Hence the distance travelled before the car stops is given by:

$s = 15t - \frac{5}{3}t^2$

$= 15(4.5) - \frac{5}{3}(4.5)^2$

$= \mathbf{33.75 \text{ m}}$

Problem 3. The distance x metres moved by a body in t seconds is given by:

$x = 3t^3 - \frac{11}{2}t^2 + 2t + 5$

Find:

(a) its velocity after t seconds;
(b) its velocity at the start and after 4 seconds,
(c) the value of t when the body comes to rest;
(d) its acceleration after t seconds;
(e) its acceleration after 2 seconds;
(f) the value of t when the acceleration is 16 m s^{-2}; and
(g) the average velocity over the third second.

(a) Distance $x = 3t^3 - \frac{11}{2}t^2 + 2t + 5$

Velocity $v = \dfrac{dx}{dt} = \mathbf{9t^2 - 11t + 2}$

(b) Velocity at the start means the velocity when $t = 0$,

i.e. $v_0 = 9(0)^2 - 11(0) + 2 = \mathbf{2 \text{ m s}^{-1}}$

Velocity after 4 seconds, $v_4 = 9(4)^2 - 11(4) + 2 = \mathbf{102 \text{ m s}^{-1}}$

(c) When the body comes to rest, $v = 0$

i.e. $9t^2 - 11t + 2 = 0$

$(9t - 2)(t - 1) = 0$

$\mathbf{t = \frac{2}{9} \text{ s or } t = 1 \text{ s}}$

(d) **Acceleration** $a = \dfrac{d^2x}{dt^2} = \mathbf{(18t - 11)}$

(e) Acceleration after 2 seconds, $a_2 = 18(2) - 11 = \mathbf{25 \text{ m s}^{-2}}$

(f) When the acceleration is 16 m s^{-2} then

$18t - 11 = 16$

$18t = 16 + 11 = 27$

$t = \frac{27}{18} = 1\frac{1}{2}$ seconds

(g) Distance travelled in the third second = (distance travelled after 3 s) − (distance travelled after 2 s)

$= [3(3)^3 - \frac{11}{2}(3)^2 + 2(3) + 5] - [3(2)^3 - \frac{11}{2}(2)^2 + 2(2) + 5]$

$= 42\frac{1}{2} - 11$

$= 31\frac{1}{2}$ m

Average velocity over the third second $= \dfrac{\text{distance travelled}}{\text{time interval}}$

$= \dfrac{31\frac{1}{2}\text{ m}}{1\text{ s}}$

$= 31\frac{1}{2}\text{ m s}^{-1}$

(Note that should a negative value occur for velocity it merely means that the body is moving in the direction opposite to that with which it started. Also if a negative value occurs for acceleration it indicates a deceleration (or a retardation).)

Further problems on velocity and acceleration may be found in Section 4, Problems 1–15, page 66.

2 Maximum and minimum values

Consider the curve shown in Fig. 2.

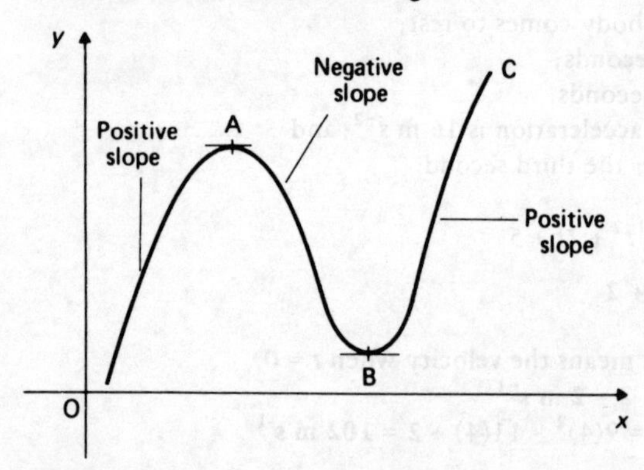

Fig. 2

The slope of the curve (i.e. $\frac{dy}{dx}$) between points O and A is positive. The slope of the curve between points A and B is negative and the slope between points B and C is again positive.

At point A the slope is zero and as x increases, the slope of the curve changes from positive just before A to negative just after. Such a point is called a **maximum value**.

At point B the slope is also zero and, as x increases, the slope of the curve changes from negative just before B to positive just after. Such a point is called a **minimum value.**

Points such as A and B are given the general name of **turning-points.**

Maximum and minimum values can be confusing inasmuch as they suggest that they are the largest and smallest values of a curve. However, by their definition this is not so. A maximum value occurs at the 'crest of a wave' and the minimum value at the 'bottom of a valley'. In Fig. 2 the point C has a larger y-ordinate value than A and point O has a smaller y ordinate than B. Points A and B are turning-points and are given the special names of maximum and minimum values respectively.

Summary

1. At a maximum point the slope $\frac{dy}{dx} = 0$ and changes from positive just before the maximum point to negative just after.
2. At a minimum point the slope $\frac{dy}{dx} = 0$ and changes from negative just before the minimum point to positive just after.

Consider the function $y = x^3 - x^2 - 5x + 6$.
The turning-points (i.e. the maximum and minimum values) may be determined without going through the tedious process of drawing up a table of values and plotting the graph.

If $\quad y = x^3 - x^2 - 5x + 6$

then $\frac{dy}{dx} = 3x^2 - 2x - 5$

Now at a maximum or minimum value $\frac{dy}{dx} = 0$.

Hence $3x^2 - 2x - 5 = 0$ for a maximum or minimum value
$(3x - 5)(x + 1) = 0$

i.e. *Either* $\quad 3x - 5 = 0$ giving $x = \frac{5}{3}$

or $\quad x + 1 = 0$ giving $x = -1$

For each value of the independent variable x there is a corresponding value of the dependent variable y.

When $x = \frac{5}{3}$, $y = [\frac{5}{3}]^3 - [\frac{5}{3}]^2 - 5[\frac{5}{3}] + 6 = -\frac{13}{27}$

When $x = -1$, $y = (-1)^3 - (-1)^2 - 5(-1) + 6 = 9$

Hence turning-points occur at $(\frac{5}{3}, -\frac{13}{27})$ and $(-1, 9)$.

The next step is to determine which of the points is a maximum and which is a minimum. There are two methods whereby this may be achieved.

Method 1

Consider firstly the point $(\frac{5}{3}, -\frac{13}{27})$.

$$\frac{dy}{dx} = 3x^2 - 2x - 5 = (3x - 5)(x + 1)$$

If x is slightly less than $\frac{5}{3}$ then $(3x - 5)$ becomes negative, $(x + 1)$ remains positive, making $\frac{dy}{dx} = (-) \times (+) = \text{negative}$

If x is slightly greater than $\frac{5}{3}$ then $(3x - 5)$ becomes positive, $(x + 1)$ remains positive, making $\frac{dy}{dx} = (+) \times (+) = \text{positive}$

Hence the slope is negative just before $(\frac{5}{3}, -\frac{13}{27})$ and positive just after. This is thus a minimum value.

Consider now the point $(-1, 9)$.

$$\frac{dy}{dx} = (3x - 5)(x + 1)$$

If x is slightly less than -1 (for example -1.1) then $(3x - 5)$ remains negative, $(x + 1)$ becomes negative, making $\frac{dy}{dx} = (-) \times (-) = \text{positive}$

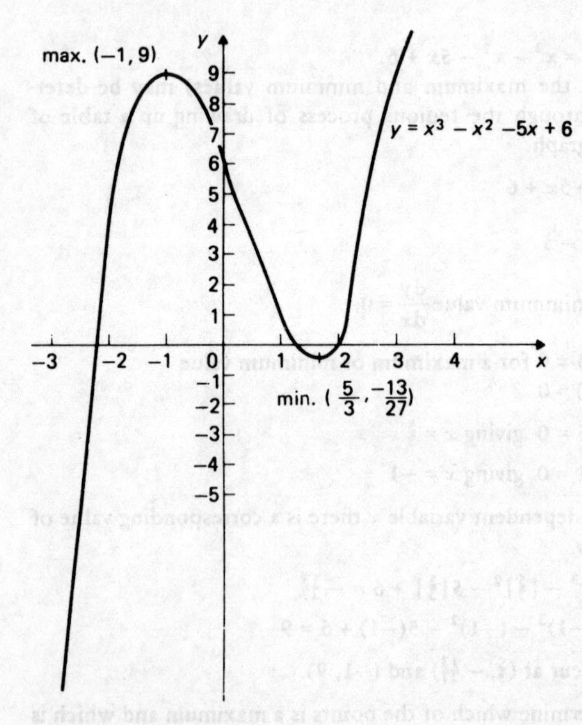

Fig. 3 Graph of $y = x^3 - x^2 - 5x + 6$

If x is slightly greater than -1 (for example -0.9) then $(3x - 5)$ remains negative, $(x + 1)$ becomes positive, making $\frac{dy}{dx} = (-) \times (+) =$ negative

Hence the slope is positive just before $(-1, 9)$ and negative just after. This is thus a **maximum value**.

Figure 3 shows a graph of $y = x^3 - x^2 - 5x + 6$ with the maximum value at $(-1, 9)$ and the minimum at $(\frac{5}{3}, -\frac{13}{27})$.

Method 2

When passing through a maximum value, $\frac{dy}{dx}$ changes from positive, through zero, to negative. By convention, moving from a positive value to a negative value is moving in a negative direction. Hence the rate of change of $\frac{dy}{dx}$ is negative.

i.e. $\frac{d}{dx}\left(\frac{dy}{dx}\right) = \frac{d^2y}{dx^2}$ **is negative at a maximum value.**

Similarly, when passing through a minimum value, $\frac{dy}{dx}$ changes from negative, through zero, to positive. By convention, moving from a negative value to a positive value is moving in a positive direction. Hence the rate of change of $\frac{dy}{dx}$ is positive.

i.e. $\frac{d^2y}{dx^2}$ **is positive at a minimum value.**

Thus, in the above example, to distinguish between the points $(\frac{5}{3}, -\frac{13}{27})$ and $(-1, 9)$ the second differential is required.

Since $\frac{dy}{dx} = 3x^2 - 2x - 5$

then $\frac{d^2y}{dx^2} = 6x - 2$

When $x = \frac{5}{3}$, $\frac{d^2y}{dx^2} = 6[\frac{5}{3}] - 2 = +8$ which is **positive**.

Hence $(\frac{5}{3}, -\frac{13}{27})$ is **a minimum point.**

When $x = -1$, $\frac{d^2y}{dx^2} = 6(-1) - 2 = -8$ which is **negative**.

Hence $(-1, 9)$ **is a maximum point.**

The actual numerical value of the second differential is insignificant for maximum and minimum values – the sign is the important factor. There are thus two methods of distinguishing between maximum and minimum values. Normally, the second method, that of determining the sign of the second differential, is preferred but sometimes the first method, that of examining the sign of the slope just before and just after the turning-point, is necessary because the second differential coefficient is too difficult to obtain.

It is possible to have a turning-point, the slope on either side of which is the same. This point is given the special name of a **point of inflexion.** At a point of inflexion $\frac{d^2y}{dx^2}$ is zero.

Maximum and minimum points and points of inflexion are given the general term of **stationary points.** Examples of each are shown in Fig. 4.

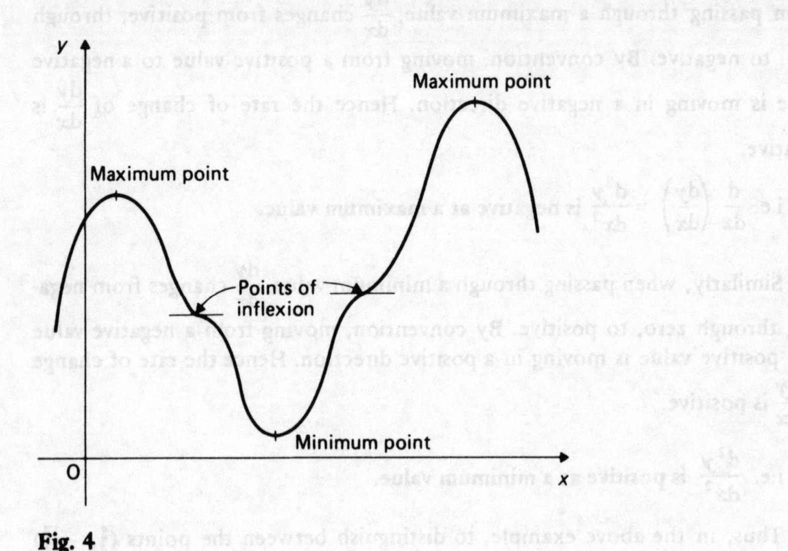

Fig. 4

Procedure for finding and distinguishing between stationary points

(i) If $y = f(x)$, find $\frac{dy}{dx}$.

(ii) Let $\frac{dy}{dx} = 0$ and solve for the value(s) of x.

(iii) Substitute the value(s) of x into the original equation, $y = f(x)$, to obtain the y-ordinate value(s). Hence the coordinates of the stationary points are established.

(iv) Find $\frac{d^2y}{dx^2}$.

or

Determine the sign of the slope of the curve just before and just after the stationary point(s).

(v) **Substitute values of x into $\frac{d^2y}{dx^2}$. If the result is:**

(a) positive – the point is a minimum value;
(b) negative – the point is a maximum value;
(c) zero – the point is a point of inflexion.

or

If the sign change for the slope of the curve is:
(a) positive to negative – the point is a maximum value;
(b) negative to positive – the point is a minimum value;
(c) positive to positive; or
(d) negative to negative – the point is a point of inflexion.

Worked problems on maximum and minimum values

Problem 1. Find the coordinates of the maximum and minimum values of the graph of $y = \frac{2x^3}{3} - 5x^2 + 12x - 7$ and distinguish between them.

From the above procedure:

(i) $y = \frac{2x^3}{3} - 5x^2 + 12x - 7$

$\frac{dy}{dx} = 2x^2 - 10x + 12$

(ii) $\frac{dy}{dx} = 0$ at a turning-point.

Therefore $2x^2 - 10x + 12 = 0$
$2(x^2 - 5x + 6) = 0$
$2(x - 2)(x - 3) = 0$
Hence $x = 2$ or $x = 3$

(iii) When $x = 2$, $y = \frac{2}{3}(2)^3 - 5(2)^2 + 12(2) - 7 = 2\frac{1}{3}$
When $x = 3$, $y = \frac{2}{3}(3)^3 - 5(3)^2 + 12(3) - 7 = 2$
The coordinates of the turning-points are thus $(2, 2\frac{1}{3})$ and $(3, 2)$.

(iv) $\frac{dy}{dx} = 2x^2 - 10x + 12$

$\frac{d^2y}{dx^2} = 4x - 10$

(v) When $x = 2$, $\frac{d^2y}{dx^2} = -2$, which is negative, giving a maximum value

When $x = 3$, $\frac{d^2y}{dx^2} = +2$, which is positive, giving a minimum value

Hence the point $(2, 2\frac{1}{3})$ is a maximum value and the point $(3, 2)$ a minimum value.

Note that with a quadratic equation there will be one turning-point. With a cubic equation (i.e. one containing a highest term of power 3) there may be two turning-points (i.e. one less than the highest power), and so on.

Problem 2. Locate the turning points on the following curves and distinguish between maximum and minimum values: (a) $x(5 - x)$; (b) $2t - e^t$; (c) $2(\theta - \ln \theta)$.

(a) Let $y = x(5 - x) = 5x - x^2$

$\frac{dy}{dx} = 5 - 2x = 0$ for a maximum or minimum value.

i.e. $x = 2\frac{1}{2}$

When $x = 2\frac{1}{2}$, $y = 2\frac{1}{2}(5 - 2\frac{1}{2}) = 6\frac{1}{4}$

Hence a turning-point occurs at $(2\frac{1}{2}, 6\frac{1}{4})$

$\frac{d^2y}{dx^2} = -2$, which is negative, giving a maximum value.

Hence $(2\frac{1}{2}, 6\frac{1}{4})$ is a maximum point.

(b) Let $y = 2t - e^t$

$\frac{dy}{dt} = 2 - e^t = 0$ for a maximum or minimum value.

i.e. $2 = e^t$

$\ln 2 = t$

$t = 0.693\,1$

When $t = 0.693\,1$, $y = 2(0.693\,1) - 2 = -0.613\,8$

Hence a turning-point occurs at $(0.693\,1, -0.613\,8)$

$\frac{d^2y}{dt^2} = -e^t$

When $t = 0.693\,1$, $\frac{d^2y}{dt^2} = -2$, which is negative, giving a maximum value.

Hence $(0.693\,1, -0.613\,8)$ is a maximum point.

(c) Let $y = 2(\theta - \ln \theta) = 2\theta - 2 \ln \theta$

$\frac{dy}{d\theta} = 2 - \frac{2}{\theta} = 0$ for a maximum or minimum value.

i.e. $\theta = 1$

When $\theta = 1$, $y = 2 - 2 \ln 1 = 2$

Hence a turning-point occurs at $(1, 2)$

$\frac{d^2y}{d\theta^2} = +\frac{2}{\theta^2}$

When $\theta = 1$, $\frac{d^2y}{d\theta^2} = +2$, which is positive, giving a minimum value.

Hence $(1, 2)$ is a minimum point.

Problem 3. Find the maximum and minimum values of the function

$$f(p) = \frac{(p - 1)(p - 6)}{(p - 10)}$$

$$f(p) = \frac{(p - 1)(p - 6)}{(p - 10)} = \frac{p^2 - 7p + 6}{(p - 10)} \text{ (i.e. a quotient)}$$

$$f'(p) = \frac{(p-10)(2p-7)-(p^2-7p+6)(1)}{(p-10)^2}$$

$$= \frac{(2p^2-27p+70)-(p^2-7p+6)}{(p-10)^2}$$

$$= \frac{p^2-20p+64}{(p-10)^2}$$

$$= \frac{(p-4)(p-16)}{(p-10)^2} = 0 \text{ for a maximum or minimum value.}$$

Therefore $(p-4)(p-16) = 0$

i.e. $p = 4$ or $p = 16$.

When $p = 4$, $f(p) = \frac{(3)(-2)}{(-6)} = 1$

When $p = 16$, $f(p) = \frac{(15)(10)}{(6)} = 25$

Hence there are turning-points at (4, 1) and (16, 25).

To use the second-derivative approach in this case would result in a complicated and long expression. Thus the slope is investigated just before and just after the turning-point.

It will be easier to use the factorised version of $f'(p)$.

i.e. $f'(p) = \frac{(p-4)(p-16)}{(p-10)^2}$

Consider the point (4, 1):

When p is just less than 4, $f'(p) = \frac{(-)(-)}{(+)}$, i.e. positive.

When p is just greater than 4, $f'(p) = \frac{(+)(-)}{(+)}$, i.e. negative.

Since the slope changes from positive to negative the point (4, 1) is a maximum.

Consider the point (16, 25):

When p is just less than 16, $f'(p) = \frac{(+)(-)}{(+)}$, i.e. negative.

When p is just greater than 16, $f'(p) = \frac{(+)(+)}{(+)}$, i.e. positive.

Since the slope changes from negative to positive the point (16, 25) is a minimum.

Since, in the question, the maximum and minimum values are asked for (and not the coordinates of the turning-points) the answers are: **maximum value = 1; minimum value = 25.**

Problem 4. Find the maximum and minimum values of $y = 1.25 \cos 2\theta + \sin \theta$ for values of θ between 0 and $\frac{\pi}{2}$ inclusive, given $\sin 2\theta = 2 \sin \theta \cos \theta$.

$$y = 1.25 \cos 2\theta + \sin \theta$$

$$\frac{dy}{d\theta} = -2.50 \sin 2\theta + \cos \theta = 0 \text{ for a maximum or minimum value.}$$

But $\sin 2\theta = 2 \sin \theta \cos \theta$

Therefore $-2.50(2 \sin \theta \cos \theta) + \cos \theta = 0$

$$-5.0 \sin \theta \cos \theta + \cos \theta = 0$$

$$\cos \theta(-5.0 \sin \theta + 1) = 0$$

Hence $\cos \theta = 0$, i.e. $\theta = 90°$ or $270°$

or $-5.0 \sin \theta + 1 = 0$, i.e. $\sin \theta = \frac{1}{5}$

$\theta = 11° \ 32'$ or $168° \ 28'$

Thus within the range $\theta = 0$ to $\theta = \frac{\pi}{2}$ inclusive, turning-points occur at $11° \ 32'$ and $90°$.

$$\frac{d^2y}{d\theta^2} = -5.0 \cos 2\theta - \sin \theta$$

When $\theta = 11° \ 32'$, $\frac{d^2y}{d\theta^2} = -4.80$, i.e. it is negative, giving a maximum value.

When $\theta = 90°$, $\frac{d^2y}{d\theta^2} = 4$, i.e. it is positive, giving a minimum value.

$y_{max} = 1.25 \cos 2(11° \ 32') + \sin (11° \ 32') = \mathbf{1.35}$

$y_{min} = 1.25 \cos 2(90°) + \sin (90°) = \mathbf{-0.25}$

Further problems on maximum and minimum values may be found in Section 4, Problems 16–41, page 67.

3 Practical problems involving maximum and minimum values

There are many practical problems on maximum and minimum values in engineering and science which can be solved using the method(s) shown in Section 2. Often the quantity whose maximum or minimum value is required appears at first to be a function of more than one variable. It is thus necessary to eliminate all but one of the variables, and this is often the only difficult part of its solution. Once the quantity has been expressed in terms of a single variable, the procedure is identical to that used in Section 2.

Worked problems on practical problems involving maximum and minimum values

---

Problem 1. A rectangular area is formed using a piece of wire 36 cm long.

Find the length and breadth of the rectangle if it is to enclose the maximum possible area.

Let the dimensions of the rectangle be x and y.
Perimeter of rectangle $= 2x + 2y = 36$

i.e. $x + y = 18$ ... (1)

Since it is the maximum area that is requured a formula for the area A must be obtained in terms of one variable only.

Area $A = xy$

From equation (1) $y = 18 - x$

Hence $A = x(18 - x) = 18x - x^2$

Now that an expression for the area has been obtained in terms of one variable it can be differentiated with respect to that variable.

$\frac{dA}{dx} = 18 - 2x = 0$ for a maximum or minimum value.

i.e. $x = 9$

$\frac{d^2A}{dx^2} = -2$, which is negative, giving a maximum value.

$y = 18 - x = 18 - 9 = 9$

Hence **the length and breadth of the rectangle of maximum area are both 9 cm**, i.e. a square gives the maximum possible area for a given perimeter length. When the perimeter of a rectangle is 36 cm the maximum area possible is 81 cm^2.

Problem 2. Find the area of the largest piece of rectangular ground that can be enclosed by 1 km of fencing if part of an existing straight wall is used as one side.

There are a large number of possible rectangular areas which can be produced from 1 000 m of fencing. Three such possibilities are shown in Fig. 5(a) where AB represents the existing wall. All three rectangles have different areas. There must be one particular condition which gives a maximum area.

Let the dimensions of any rectangle be x and y as shown in Fig. 5 (b).

Then $2x + y = 1\,000$ ... (1)

Area of rectangle, $A = xy$ ... (2)

Since it is the **maximum area** that is required, a formula for the area A must be obtained in terms of one variable only.

From equation (1) $y = 1\,000 - 2x$

Hence $A = x(1\,000 - 2x) = 1\,000x - 2x^2$

$\frac{dA}{dx} = 1\,000 - 4x = 0$ for a maximum or minimum value

i.e. $x = 250$

$$\frac{d^2A}{dx^2} = -4, \text{ which is negative, giving a maximum value.}$$

When $x = 250$, $y = 1\,000 - 2(250) = 500$

Hence the maximum possible area $= xy\ \text{m}^2 = (250)(500)\ \text{m}^2 = 125\,000\ \text{m}^2$.

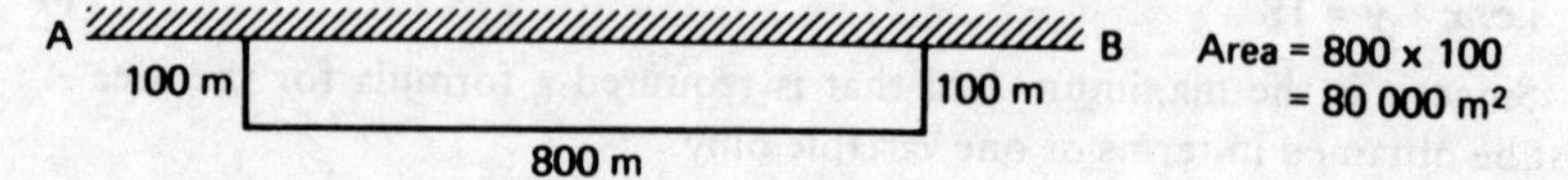

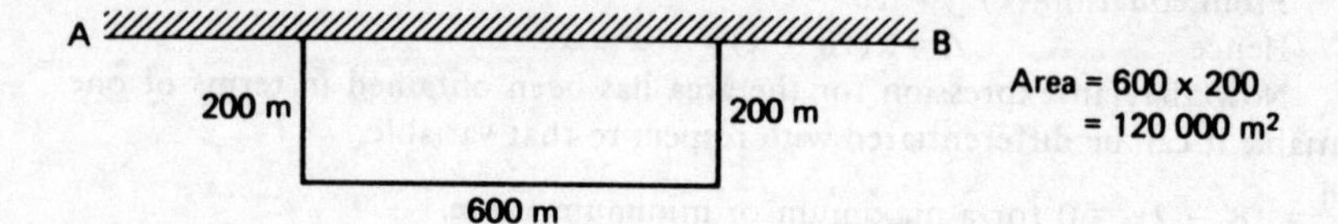

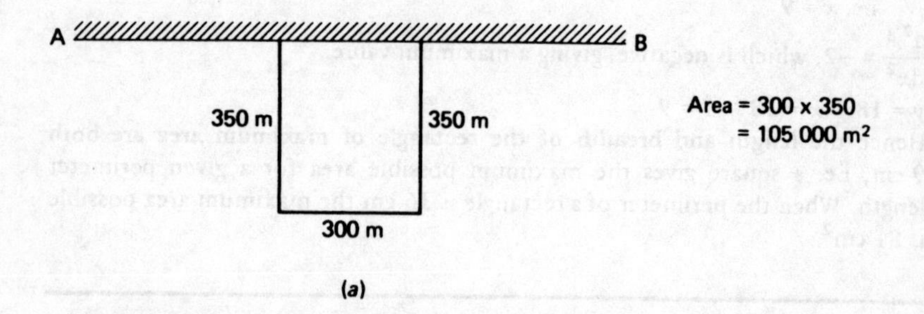

(*a*)

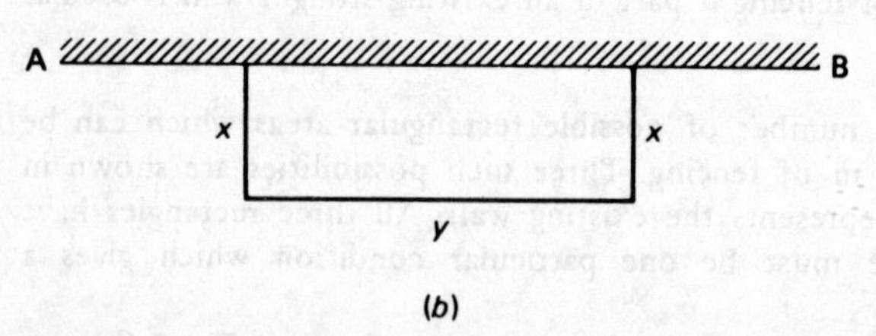

(*b*)

Fig. 5

Problem 3. A lidless, rectangular box with square ends is to be made from a thin sheet of metal. What is the least area of the metal for which the volume is $4\frac{1}{2}$ m^3?

Let the dimensions of the box be x metres by x metres by y metres.

Volume of box $= x^2y = 4\frac{1}{2}$ . . . (1)

Surface area A of box consists of: two ends $= 2x^2$

two sides $= 2xy$

base $= xy$

$A = 2x^2 + 2xy + xy = 2x^2 + 3xy$ ... (2)

Since it is the least (i.e. minimum in this case) area that is required, a formula for the area A must be obtained in terms of one variable only.

From equation (1), $y = \frac{4\frac{1}{2}}{x^2} = \frac{9}{2x^2}$

Substituting $y = \frac{9}{2x^2}$ in equation (2) gives:

$$A = 2x^2 + 3x\left(\frac{9}{2x^2}\right) = 2x^2 + \frac{27}{2x}$$

$$\frac{dA}{dx} = 4x - \frac{27}{2x^2} = 0 \text{ for a maximum or minimum value}$$

$$4x = \frac{27}{2x^2}$$

$$x^3 = \frac{27}{8}, \text{ i.e. } x = \frac{3}{2}$$

$$\frac{d^2A}{dx^2} = 4 + \frac{27}{x^3}$$

When $x = \frac{3}{2}$, $\frac{d^2A}{dx^2} = 4 + \frac{27}{(\frac{3}{2})^3} = +12$, which is positive, giving a minimum (or least) value.

When $x = \frac{3}{2}$, $y = \frac{9}{2x^2} = \frac{9}{2(\frac{3}{2})^2} = 2$

Therefore area $A = 2x^2 + 3xy = 2[\frac{3}{2}]^2 + 3[\frac{3}{2}](2) = 13\frac{1}{2}$

Hence the least possible area of metal required to form a rectangular box with square ends of volume $4\frac{1}{2}$ m^3 is $13\frac{1}{2}$ m^2.

Problem 4. Find the base radius and height of the cylinder of maximum volume which can be cut from a sphere of radius 10.0 cm.

A cylinder of radius r and height h is shown in Fig. 6 enclosed in a sphere of radius R = 10.0 cm.

Volume of cylinder, $V = \pi r^2 h$ ... (1)

Using the theorem of Pythagoras on the triangle ABC of Fig. 6 gives:

$$r^2 + \left[\frac{h}{2}\right]^2 = R^2$$

i.e. $r^2 + \frac{h^2}{4} = 100$ ... (2)

Since it is the maximum volume that is required, a formula for the volume V must be obtained in terms of one variable only.

From equation (2), $r^2 = 100 - \frac{h^2}{4}$

Substituting $r^2 = 100 - \frac{h^2}{4}$ in equation (1) gives:

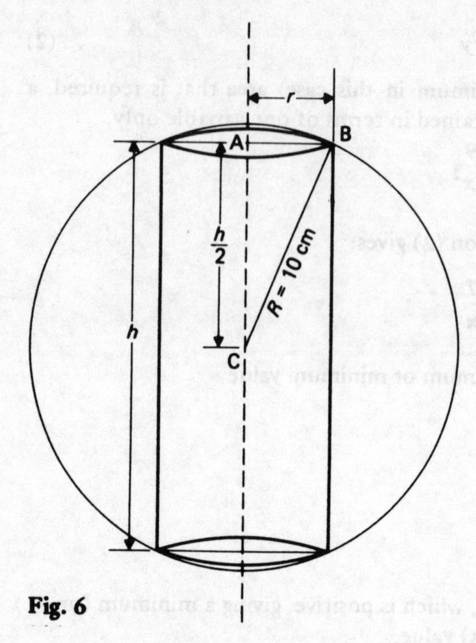

Fig. 6

$$V = \pi\left[100 - \frac{h^2}{4}\right]h = 100\pi\, h - \frac{\pi h^3}{4}$$

$\dfrac{dV}{dh} = 100\pi - \dfrac{3\pi}{4}h^2 = 0$ for a maximum or minimum value.

$$100\pi = \frac{3\pi}{4}h^2$$

$$h^2 = \frac{400}{3}$$

$h = 11.55$ cm ($h = -11.55$ cm is neglected for obvious reasons)

$$\frac{d^2V}{dh^2} = -\tfrac{3}{2}\pi h$$

When $h = 11.55$, $\dfrac{d^2V}{dh^2} = -\tfrac{3}{2}\pi(11.55) = -54.43$ which is negative, giving a maximum value.

From equation (2) $r^2 = 100 - \dfrac{h^2}{4}$

$$r = \sqrt{\left(100 - \frac{h^2}{4}\right)} = 8.164 \text{ cm}$$

Hence the cylinder having the largest volume that can be cut from a sphere of radius 10.0 cm is one in which the base radius is 8.164 cm and the height is 11.55 cm.

Problem 5. A piece of wire 4.0 m long is cut into two parts one of which is bent into a square and the other bent into a circle. Find the radius of the circle if the sum of their areas is a minimum.

Let the square be of side x m and the circle of radius r m.
The sum of the perimeters of the square and circle is given by:

$$4x + 2\pi r = 4$$

or $2x + \pi r = 2$ ... (1)

Total area A of the two shapes, $A = x^2 + \pi r^2$ ... (2)

Since it is the **minimum area** that is required a formula for the area A must be obtained in terms of one variable only.

From equation (1), $x = \dfrac{2 - \pi r}{2}$

Substituting $x = \dfrac{2 - \pi r}{2}$ in equation (2) gives:

$$A = \left(\frac{2 - \pi r}{2}\right)^2 + \pi r^2 = \frac{4 - 4\pi r + \pi^2 r^2}{4} + \pi r^2$$

i.e. $A = 1 - \pi r + \dfrac{\pi^2 r^2}{4} + \pi r^2$

$\dfrac{dA}{dr} = -\pi + \dfrac{\pi^2 r}{2} + 2\pi r = 0$ for a maximum or minimum value.

i.e. $\pi = r\left[\dfrac{\pi^2}{2} + 2\pi\right]$

$$r = \frac{\pi}{\left(\dfrac{\pi^2}{2} + 2\pi\right)} = \frac{1}{\left(\dfrac{\pi}{2} + 2\right)} = 0.280 \text{ m}$$

$\dfrac{d^2A}{dr^2} = \dfrac{\pi^2}{2} + 2\pi = 11.22$, which is positive, giving a minimum value.

Hence for the sum of the areas of the square and circle to be a minimum the radius of the circle must be 28.0 cm.

Problem 6. Find the base radius of a cylinder of maximum volume which can be cut from a cone of height 12 cm and base radius 9 cm.

A cylinder of base radius r cm and height h cm is shown enclosed in a cone of height 12 cm and base radius 9 cm in Fig. 7.

Volume of the cylinder, $V = \pi r^2 h$ ... (1)

By similar triangles: $\dfrac{12 - h}{r} = \dfrac{12}{9}$ ... (2)

Since it is the **maximum volume** that is required a formula for the volume V must be obtained in terms of one variable only.

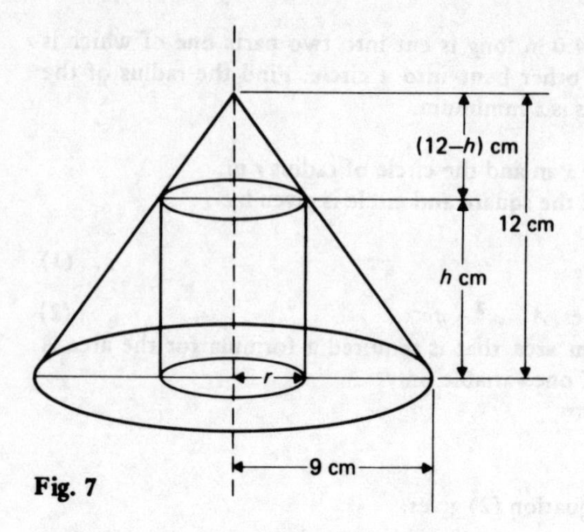

Fig. 7

From equation (2), $9(12 - h) = 12r$

$$108 - 9h = 12r$$

$$h = \frac{108 - 12r}{9}$$

Substituting for h in equation (1) gives:

$$V = \pi r^2\left[\frac{108 - 12r}{9}\right] = 12\pi r^2 - \frac{4\pi r^3}{3}$$

$$\frac{dV}{dr} = 24\pi r - 4\pi r^2 = 0 \text{ for a maximum or minimum value.}$$

i.e. $4\pi r(6 - r) = 0$

Therefore $r = 0$ or $r = 6$.

$$\frac{d^2V}{dr^2} = 24\pi - 8\pi r$$

When $r = 0$, $\frac{d^2V}{dr^2}$ is positive, giving a minimum value (which we would expect).

When $r = 6$, $\frac{d^2V}{dr^2} = 24\pi - 48\pi = -24\pi$, which is negative, giving a maximum value.

Hence a cylinder of maximum volume having a base radius of 6 cm can be cut from a cone of height 12 cm and base radius 9 cm.

Problem 7. A rectangular sheet of metal which measures 24.0 cm by 16.0 cm has squares removed from each of the four corners so that an open box may be formed. Find the maximum possible volume for the box.

The squares which are to be removed are shown shaded and having side x cm in Fig. 8.

To form a box the metal has to be bent upwards along the broken lines.

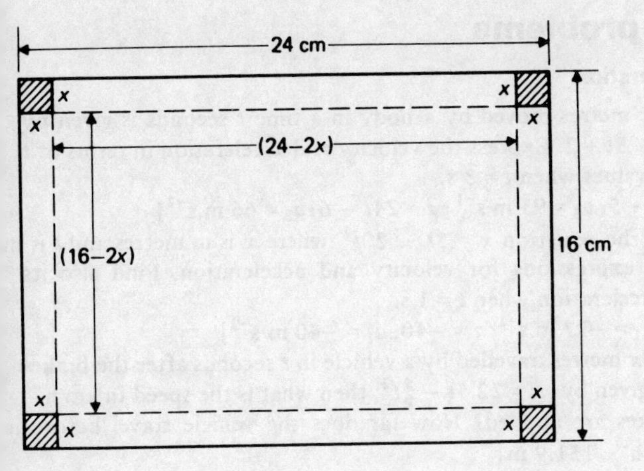

Fig. 8

The dimensions of the box will be: length = $(24.0 - 2x)$ cm
breadth = $(16.0 - 2x)$ cm
height = x cm

If volume of box is V cm^3, $V = (24.0 - 2x)(16.0 - 2x)(x)$
$= 384x - 80x^2 + 4x^3$

$$\frac{dV}{dx} = 384 - 160x + 12x^2 = 0 \text{ for a maximum or minimum value.}$$

i.e. $4(3x^2 - 40x + 96) = 0$

$$x = \frac{40 \pm \sqrt{[(-40)^2 - 4(3)(96)]}}{6}$$

$x = 10.194$ cm or $x = 3.139$ cm

Since the breadth $= (16.0 - 2x)$ cm, $x = 10.194$ cm is an impossible solution to this problem and is thus neglected.

Hence $x = 3.139$ cm

$$\frac{d^2V}{dx^2} = -160 + 24x$$

When $x = 3.139$ cm, $\frac{d^2V}{dx^2} = -160 + 24(3.139) = -84.66$ which is negative, giving a maximum value.

The dimensions of the box are: length = $24.0 - 2(3.139) = 17.72$ cm
breadth = $16.0 - 2(3.139) = 9.722$ cm
height = 3.139 cm

Maximum volume = $(17.72)(9.722)(3.139) = 540.8$ cm^3.

Further problems on practical maximum and minimum problems may be found in the following section (4), Problems 42–69, page 68. There are also some further typical practical differentiation examples, Problems 70–81, page 71.

4 Further problems

Velocity and acceleration

1. The distance x metres moved by a body in a time t seconds is given by $x = 4t^3 - 3t^2 + 5t + 2$. Express the velocity and acceleration in terms of t and find their values when $t = 3$ s.
[$v = 12t^2 - 6t + 5$; $v_3 = 95$ m s^{-1}; $a = 24t - 6$; $a_3 = 66$ m s^{-2}]
2. A body obeys the equation $x = 3t - 20t^2$ where x is in metres and t is in seconds. Find expressions for velocity and acceleration. Find also its velocity and acceleration when $t = 1$ s.
[$v = 3 - 40t$; $v_1 = -37$ m s^{-1}; $a = -40$; $a_1 = -40$ m s^{-2}]
3. If the distance x metres travelled by a vehicle in t seconds after the brakes are applied is given by: $x = 22.5t - \frac{5}{6}t^2$, then what is the speed in km h^{-1} when the brakes are applied? How far does the vehicle travel before it stops? [81 km h^{-1}; 151.9 m]
4. An object moves in a straight line so that after t seconds its distance x metres from a fixed point on the line is given by $x = \frac{2}{3}t^3 - 5t^2 + 8t - 6$. Obtain an expression for the velocity and acceleration of the object after t seconds and hence calculate the values of t when the object is at rest.
[$v = 2t^2 - 10t + 8$; $a = 4t - 10$; $t = 1$ s or 4 s]

In Problems 5–9, x denotes the distance in metres of a body moving in a straight line, from a fixed point on the line and t denotes the time in seconds measured from a certain instant. Find the velocity and acceleration of the body when t has the given values. Find also the values of t when the body is momentarily at rest.

5. $x = \frac{4}{3}t^3 - 4t^2 + 3t - 2$; $t = 2$. [3 m s^{-1}; 8 m s^{-2}; $t = \frac{1}{2}$ or $1\frac{1}{2}$ s]
6. $x = 3\cos 2t$; $t = \frac{\pi}{4}$. [-6 m s^{-1}; 0; $t = 0, \frac{\pi}{2}, \pi, \frac{3\pi}{2}$, etc.]
7. $x = t^4 - \frac{1}{2}t^2 + 1$; $t = 1$. [3 m s^{-1}; 11 m s^{-2}; $t = 0$ or $\pm\frac{1}{2}$ s]
8. $x = 4t + 2\cos 2t$; $t = 0$. [4 m s^{-1}; -8 m s^{-2}; $t = \frac{\pi}{4}, \frac{5\pi}{4}, \frac{9\pi}{4}, \frac{13\pi}{4}$, etc.]
9. $x = \frac{t^4}{4} - \frac{5}{3}t^3 + 3t^2 + 5$; $t = 2$. [0; -2 m s^{-2}; $t = 0$, 2 or 3 s]
10. The distance s metres moved by a point in t seconds is given by $s = 5t^3 + 4t^2 - 3t + 2$. Find:
(a) expressions for velocity and acceleration in terms of t;
(b) the velocity and acceleration after 3 seconds; and
(c) the average velocity over the fourth second.
(a) [$v = (15t^2 + 8t - 3)$ m s^{-1}; $a = (30t + 8)$ m s^{-2}]
(b) [156 m s^{-1}; 98 m s^{-2}] (c) [210 m s^{-1}]
11. The distance x metres moved by a body in t seconds is given by $x = \frac{16}{3}t^3 - 32t^2 + 39t - 16$. Find:
(a) the velocity and acceleration at the start;
(b) the velocity and acceleration at $t = 3$ seconds;
(c) the values of t when the body is at rest;

(d) the value of t when the acceleration is 16 m s^{-2}; and
(e) the distance travelled in the second second.
(a) [39 m s^{-1}; −64 m s^{-2}] (b) [−9 m s^{-1}; 32 m s^{-2}] (c) [$t = \frac{3}{4}$ or $3\frac{1}{4}$ s]
(d) [$2\frac{1}{2}$ s] (e) [$-19\frac{2}{3}$ m (i.e. in the opposite direction to that in which the body initially moved)]

12. The displacement y centimetres of the slide valve of an engine is given by the expression $y = 2.6 \cos 5\pi t + 3.8 \sin 5\pi t$. Find an expression for the velocity v of the valve and evaluate the velocity (in metres per second) when $t = 20$ ms. [$v = 5\pi(3.8 \cos 5\pi t - 2.6 \sin 5\pi t)$; 0.442 m s^{-1}]

13. At any time t seconds the distance x metres of a particle moving in a straight line from a fixed point is given by: $x = 5t + \ln(1 - 2t)$. Find:
(a) expressions for the velocity and acceleration in terms of t;
(b) the initial velocity and acceleration;
(c) the velocity and acceleration after 2 s; and
(d) the time when the velocity is zero.
(a) $\left[\left(5 - \frac{2}{(1-2t)}\right) \text{m s}^{-1}; \left(\frac{-4}{(1-2t)^2}\right) \text{m s}^{-2}\right]$
(b) [3 m s^{-1}; −4 m s^{-2}] (c) [$5\frac{2}{3}$ m s^{-1}; $-\frac{4}{9}$ m s^{-2}] (d) [$\frac{3}{10}$ s]

14. If the equation $\theta = 12\pi + 27t - 3t^2$ gives the angle in radians through which a wheel turns in t seconds, find how many seconds the wheel takes to come to rest. Calculate the angle turned through in the last second of movement. [$4\frac{1}{2}$ s; 3 radians]

15. A missile fired from ground level rises s metres in t seconds, and $s = 75t - 12.5t^2$. Determine:
(a) the initial velocity of the missile;
(b) the time when the height of the missile is a maximum;
(c) the maximum height reached; and
(d) the velocity with which the missile strikes the ground.
(a) [75 m s^{-1}] (b) [3 s] (c) [112.5 m] (d) [75 ms^{-1}]

Maximum and minimum values

In Problems 16–20 find the turning-points and distinguish between them by examining the sign of the slope on either side.

16. $y = 2x^2 - 4x$ [min. (1, −2)]
17. $y = 3t^2 - 2t + 6$ [min. ($\frac{1}{3}$, $5\frac{2}{3}$)]
18. $x = \theta^3 - 3\theta + 3$ [max. (−1, 5); min. (1, 1)]
19. $y = 3x^3 + 6x^2 + 3x - 2$ [max. (−1, −2); min. ($-\frac{1}{3}$, $-2\frac{4}{9}$)]
20. $y = 7t^3 - 4t^2 - 5t + 6$ [max. ($-\frac{1}{3}$, $6\frac{26}{27}$); min. ($\frac{5}{7}$, $2\frac{46}{49}$)]

Locate the turning-points on the curves in Problems 21–39 and determine whether they are maximum and minimum points.

21. $y = x(7 - x)$ [max. ($3\frac{1}{2}$, $12\frac{1}{4}$)]
22. $y = 4x^2 - 2x + 3$ [min. ($\frac{1}{4}$, $2\frac{3}{4}$)]
23. $y = 2x^3 + 7x^2 + 4x - 3$ [max. (−2, 1); min. ($-\frac{1}{3}$, $-3\frac{17}{27}$)]
24. $2pq = 18p^2 + 8$ [max. ($-\frac{2}{3}$, −12); min. ($\frac{2}{3}$, 12)]
25. $y = 3t + e^{-t}$ [min. (−1.098 6, −0.2958)]
26. $x = 3 \ln \theta - 4\theta$ [max. (0.75, −3.863)]

27. $S = 5t^3 - \frac{3}{2}t^2 - 12t + 6$ [max. $(-\frac{4}{5}, 12\frac{2}{25})$; min. $(1, -2\frac{1}{2})$]
28. $y = 4x - 2 \ln x$ [min. (0.5, 3.386)]
29. $y = 3x - e^x$ [max. (1.098 6, 0.295 8)]
30. $p = \dfrac{(q-1)(q-3)}{q}$ [max. (−1.732, −7.464); min. (1.732, −0.535 9)]
31. $y = \dfrac{(x-2)(x-5)}{(x-6)}$ [max. (4, 1); min. (8, 9)]
32. $y = 3 \sin \theta - 4 \cos \theta$ in the range θ to 2π. [max. 5 at 143° 8′; min. −5 at 323° 8′]
33. $y = 4 \cos 2\theta + 3 \sin \theta$ in the range 0 to $\dfrac{\pi}{2}$ inclusive, given that $\sin 2\theta = 2 \sin \theta \cos \theta$. [max. 4.281 2 at 10° 48′; min. −1 at 90°]
34. $V = l^2(l-1)$ [max. (0, 0); min. $(\frac{2}{3}, \frac{-4}{27})$]
35. $y = 8x + \dfrac{1}{2x^2}$ [min. $(\frac{1}{2}, 6)$]
36. $x = t^3 + \dfrac{t^2}{2} - 2t + 4$ [max. $-1, 5\frac{1}{2}$); min. $(\frac{2}{3}, 3\frac{5}{27})$]
37. $y = \dfrac{3x}{(x-1)(x-4)}$ [max. (2, −3); min. $(-2, -\frac{1}{3})$]
38. $y = (x-1)^3 + 3x(x-2)$ [max. (−1, 1); min. (1, −3)]
39. $y = \frac{1}{2} \ln (\sin x) - \sin x$ in the range 0 to $\dfrac{\pi}{4}$ [max. −0.846 6 at 30°]
40. (a) If $p + q = 7$, find the maximum value of $3pq + q^2$.
    (b) If $3a - 2b = 5$, find the least value of $2a^2b$.
    (a) [$55\frac{1}{8}$] (b) [$-2\frac{14}{243}$ or −2.057 6]
41. The sum of a number and its reciprocal is to be a minimum. Find the number. [1]

Practical maximum and minimum problems

42. Find the maximum area of a rectangular piece of ground that can be enclosed by 200 m of fencing. [2 500 m^2]
43. A rectangular area is formed using a piece of wire of length 26 cm. Find the dimensions of the rectangle if it is to enclose the maximum possible area. [$6\frac{1}{2}$ cm by $6\frac{1}{2}$ cm]
44. A shell is projected upwards with a speed of 12 m s^{-1} and the distance vertically s metres is given by $s = 12t - 3t^2$, where t is the time in seconds. Find the maximum height reached. [12 m]
45. A length of 42 cm of thin wire is bent into a rectangular shape with one side repeated. Find the largest area that can be enclosed. [73.5 cm^2]
46. Find the area of the largest piece of rectangular ground that can be enclosed by 800 m of fencing if part of an existing wall is used on one side. [80 000 m^2]
47. The bending moment M of a beam of length l at a distance a from one end is given by $M = \dfrac{Wa}{2}(l - a)$, where W is the load per unit length. Find

the maximum bending moment. $\left[\frac{Wl^2}{8}\right]$

48. A lidless box with square ends is to be made from a thin sheet of metal. What is the least area of the metal for which the volume of the box is 6.64 m^3? [17.50 m^2]
49. Find the height and the radius of a cylinder of volume 150 cm^3 which has the least surface area. [5.759 cm; 2.879 cm]
50. Find the height of a right circular cylinder of greatest volume which can be cut from a sphere of radius R. $\left[\frac{2R}{\sqrt{3}}\right]$
51. The power P developed in a resistor R by a battery of e.m.f. E and internal resistance r is given by $P = \frac{E^2 R}{(R+r)^2}$.

    Differentiate P with respect to R and show that the power is a maximum when $R = r$.
52. A piece of wire 5.0 m long is cut into two parts, one of which is bent into a square and the other into a circle. Find the diameter of the circle if the sum of their areas is a minimum. [0.700 m]
53. Find the height of a cylinder of maximum volume which can be cut from a cone of height 15 cm and base radius 7.5 cm. [5 cm]
54. An alternating current is given by $i = 100 \sin(50\pi t + 0.32)$ amperes, where t is the time in seconds. Determine the maximum value of the current and the time when this maximum first occurs. [100 amperes when t = 7.96 ms].
55. A frame for a box kite with a square cross-section is made of 16 pieces of wood as shown in Fig. 9. Find the maximum volume of the frame if a total length of 12 m of wood is used. [$\frac{4}{9}$ m^3]

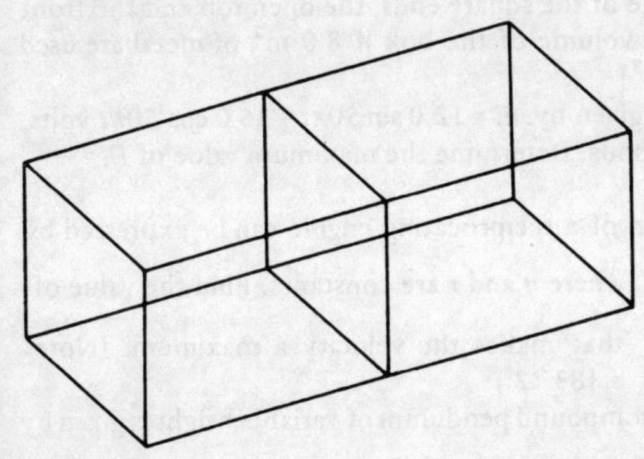

Fig. 9

56. A rectangular box with a lid which covers the top and front has a volume of 150 cm^3 and the length of the base is to be $1\frac{1}{2}$ times the height. Find the dimensions of the box so that the surface area shall be a minimum. [3.816 cm by 5.724 cm by 6.867 cm]

57. The force F required to move a body along a rough horizontal plane is given by $F = \dfrac{\mu W}{\cos\theta + \mu\sin\theta}$ where μ is the coefficient of friction and θ the angle to the direction of F. If F varies with θ show that F is a minimum when $\tan\theta = \mu$.
58. A closed cylindrical container has a surface area of 300 cm^2. Find its dimensions for maximum volume.
[radius = 3.989 cm; height = 7.981 cm]
59. A rectangular block of metal, with a square cross-section, has a total surface area of 240 cm^2. Find the maximum volume of the block of metal. [253.0 cm^3]
60. The displacement s metres in a damped harmonic oscillation is given by $s = 4e^{-2t}\sin 2t$, where t is the time in milliseconds. Find the values of t to give maximum displacements. $\left[\frac{\pi}{8}, \frac{5\pi}{8}, \frac{9\pi}{8}, \text{and so on}\right]$
61. A square sheet of metal of side 25.0 cm has squares cut from each corner, so that an open box may be formed. Find the surface area and the volume of the box if the volume is to be a maximum.
[555.6 cm^2; 1 157 cm^3]
62. A right circular cylinder of maximum volume is to be cut from a sphere of radius 14.0 cm. Determine the base diameter and the height of the cylinder. [22.86 cm; 16.17 cm]
63. The speed v of a signal transmitted through a cable is given by $v = kx^2 \ln\frac{1}{x}$, where x is the ratio of the inner to the outer diameters of the core and k is a constant. Find the value of x for maximum speed of the transmitted signal. [$x = e^{-\frac{1}{2}} = 0.606\ 5$]
64. An open rectangular box with square ends is fitted with an overlapping lid which covers the whole of the square ends, the open top and the front face. Find the maximum volume of the box if 8.0 m^2 of metal are used altogether. [0.871 m^3]
65. An electrical voltage E is given by: $E = 12.0\sin 50\pi t + 36.0\cos 50\pi t$ volts, where t is the time in seconds. Determine the maximum value of E.
[37.95 volts]
66. The velocity v of a piston of a reciprocating engine can be expressed by $v = 2\pi nr\left(\frac{\sin 2\theta}{16} + \sin\theta\right)$, where n and r are constants. Find the value of θ between 0° and 360° that makes the velocity a maximum. (Note: $\cos 2\theta = 2\cos^2\theta - 1$.) [83° 2′]
67. The periodic time T of a compound pendulum of variable height is given by $T = 2\pi\sqrt{\left[\frac{h^2 + k^2}{gh}\right]}$, where k and g are constants. Find the minimum value of T. $\left[T_{min} = 2\pi\sqrt{\left(\frac{2k}{g}\right)} \text{ when } h = k\right]$
68. The heat capacity (C) of carbon monoxide varies with absolute tempera-

ture (T) as shown: $C = 26.53 + 7.70 \times 10^{-3}T - 1.17 \times 10^{-6}T^2$. Determine the maximum value of C and the temperature at which it occurs. [C = 39.20, $T = 3.291 \times 10^3$]

69. The electromotive force (E) of the Clark cell is given by $E = 1.4 - 0.001\,2\,(T - 288) - 0.000\,007\,(T - 288)^2$ volts. Determine the maximum value of E. [1.451 4 volts]

Practical differentiation

70. The length l metres of a certain rod at temperature t°C is given by $l = 1 + 0.000\,02t + 0.000\,000\,2t^2$. Find the rate at which l increases with respect to t $\left(\text{i.e. } \frac{dl}{dt}\right)$ when the temperature is: (a) 100°C; and (b) 300°C. (a) [0.000 06 m °C^{-1}] (b) [0.000 14 m °C^{-1}]

71. An alternating voltage v volts is given by $v = 125 \sin 80\,t$, where t is the time in seconds. Calculate the rate of change of voltage $\left(\text{i.e. } \frac{dv}{dt}\right)$ when t = 20 ms. [−292 volts per second]

72. In a first-order reaction the concentration c after time t is governed by the relation $c = ae^{-kt}$, where a is the initial concentration and k is the rate constant. Show that the value of $\frac{dc}{dt}$ is given by $-kc$.

73. A displacement s metres is given by $s = \frac{1}{3}t^3 - 4\,t^2 + 15\,t$, where t is the time in seconds. At what times is the velocity v $\left(\text{i.e. } \frac{ds}{dt}\right)$ zero? Determine the acceleration $\left(\text{i.e. } \frac{dv}{dt}\right)$ when t = 6 seconds. [t = 3 s or 5 s; 4 m s^{-2}]

74. The luminous intensity, I candelas, of a lamp at different voltages V is given by $I = 4 \times 10^{-4}\,V^2$. Find the voltage at which the light is increasing at a rate of 0.2 candelas per volt, $\left(\text{i.e. when } \frac{dI}{dV} = 0.2\right)$ [250 V]

75. The relationship between pressure p and volume v is given by $pv^n = k$, where n and k are constants. Prove that $v^{n+1}\,\frac{dp}{dv} + kn = 0$.

76. An alternating current i amperes is given by $i = 70 \sin 2\pi f\,t$, where f is the frequency in hertz and t the time in seconds. Find the rate of change of current $\left(\text{i.e. } \frac{di}{dt}\right)$ when t = 20 ms, given that the frequency is 50 hertz. [7 000π A s^{-1}]

77. Newton's law of cooling is given by $\theta = \theta_0 e^{-kt}$, where the excess of temperature at zero time is θ_0°C and at time t seconds is θ°C. Determine the rate of change of temperature $\left(\text{i.e. } \frac{d\theta}{dt}\right)$, given that θ_0 = 15°C, $k = -0.017$ and t = 60 seconds. [0.707 °C s^{-1}]

78. A coil has a self inductance L of 2 henries and a resistance R of 100 ohms. A d.c. supply voltage E of 100 volts is applied to the coil. The

instantaneous current i amperes is given by $i = \frac{E}{R}\left(1 - e^{-\frac{Rt}{L}}\right)$. Find: (a) the rate at which the current increases at the moment of switching on $\left(\text{i.e. } \frac{di}{dt} \text{ when } t = 0\right)$; and (b) the rate after 10 ms. (a) [50 A s^{-1}] (b) [30.3 A s^{-1}]

79. A fully charged capacitor C of 0.1 microfarads has a potential difference V of 150 volts between its plates. The capacitor is then discharged through a resistor R of 1 megohm. If the potential difference v across the plate at any time t seconds after closing the circuit is given by $v = Ve^{-\frac{t}{RC}}$ calculate:

(a) the initial rate of loss of voltage $\left(\text{i.e. } \frac{dv}{dt} \text{ at } t = 0\right)$;

(b) the rate after 0.1 seconds.

(a) [1 500 V s^{-1}] (b) [552 V s^{-1}]

80. The pressure p of the atmosphere at height h above ground level is given by $p = p_0 e^{-\frac{h}{c}}$, where p_0 is the pressure at ground level and c is a constant.

Determine the rate of change of pressure with height $\left(\text{i.e. } \frac{dp}{dh}\right)$ when p_0 is 1.013×10^5 pascals and c is 6.062×10^4 at 1 500 metres. [−1.630 Pa/m]

81. The displacement s cm of the end of a stiff spring at time t seconds is given by $s = ae^{-kt} \sin 2\pi f t$. Find the velocity v $\left(\text{i.e. } \frac{ds}{dt}\right)$ and acceleration a $\left(\text{i.e. } \frac{d^2s}{dt^2}\right)$ of the end of the spring after 3 seconds if $a = 4$, $k = 0.8$ and $f = 2$. [4.56 cm s^{-1}, −7.30 cm s^{-2}]

Chapter 4

Methods of integration

1 Introduction to integration

The process of integration reverses the process of differentiation. In differentiation, if $f(x) = x^2$ then $f'(x) = 2x$. Since integration reverses the process of moving from $f(x)$ to $f'(x)$, it follows that the integral of $2x$ is x^2, i.e. it is the process of moving from $f'(x)$ to $f(x)$. Similarly, if $y = x^3$ then $\frac{dy}{dx} = 3x^2$. Reversing this process shows that the integral of $3x^2$ is x^3.

Integration is also a process of summation or adding parts together and an elongated 'S', shown as $\int$, is used to replace the words 'the integral of'.

Thus $\int 2x = x^2$ and $\int 3x^2 = x^3$.

In differentiation, the differential coefficient $\frac{dy}{dx}$ or $\frac{d}{dx}[f(x)]$ indicates that a function of x is being differentiated with respect to x, the dx indicating this. In integration, the variable of integration is shown by adding d (the variable) after the function to be integrated. Thus $\int 2x\, dx$ means 'the integral of $2x$ with respect to x' and $\int 3u^2\, du$ means 'the integral of $3u^2$ with respect to u'. It follows that $\int y\, dx$ means 'the integral of y with respect to x' and since only functions of x can be integrated with respect to x, y must be expressed as a function of x before the process of integration can be performed.

The arbitrary constant of integration

The differential coefficient of x^2 is $2x$, hence $\int 2x\ dx = x^2$. Also, the differential coefficient of $x^2 + 3$ is $2x$, hence $\int 2x\ dx = x^2 + 3$. Since the differential coefficient of any constant is zero, it follows that the differential coefficient of $x^2 + c$, where c is any constant, is $2x$. To allow for the possible presence of this constant, whenever the process of integration is performed the constant should be added to the result. Hence $\int 2x\ dx = x^2 + c$.

c is called the arbitrary constant of integration and it is important to include it in all work involving the process of determining integrals. Its omission will result in obtaining incorrect solutions in later work, such as in the solution of differential equations (see chapter 7).

2 The general solution of integrals of the form x^n

$$\int x^n\ dx = \frac{x^{n+1}}{n+1} + c$$

In order to integrate x^n it is necessary to:

(a) increase the power of x by 1, i.e. the power of x^n is raised by 1 to x^{n+1};
(b) divide by the new power of x, i.e. x^{n+1} is divided by $n + 1$; and
(c) add the arbitrary constant of integration, c.

Thus to integrate x^4, the power of x is increased by 1 to x^{n+1} or x^{4+1}, i.e. x^5 and the term is divided by $(n + 1)$ or $(4 + 1)$, i.e. 5.

So the integral of x^4 is $\frac{x^5}{5} + c$.

In the general solution of $\int x^n\ dx$ given above, n may be a positive or negative integer or fraction, or zero, with just one exception, that being $n = -1$.

It was shown in differentiation that $\frac{d}{dx}(\ln x) = \frac{1}{x}$

Thus $\int \frac{1}{x}\ dx\ (= \int x^{-1}\ dx) = \ln x + c$

More generally, $\frac{d}{dx}(\ln ax) = \frac{d}{dx}(\ln x + \ln a)$

$$= \frac{1}{x} + 0$$

Therefore, $\int \frac{1}{x}\ dx = \ln ax + c$

Rules of integration

Three of the basic rules of integration are:

(i) The integral of a constant k is $kx + c$. For example,

$\int 5\,dx = \int 5x^0\,dx$ since $x^0 = 1$:

Applying the standard integral $\int x^n\,dx = \frac{x^{n+1}}{n+1} + c$ gives:

$$\int 5\,dx = \frac{5x^{0+1}}{0+1} + c = 5x + c$$

(ii) As in differentiation, constants associated with variables are carried forward, i.e. they are not involved in the integration. For example,

$$\int 3x^4\,dx = 3\int x^4\,dx = 3\left(\frac{x^5}{5}\right) + c$$
$$= \frac{3}{5}x^5 + c$$

(iii) As in differentiation, the rules of algebra apply where functions of a variable are added or subtracted. For example,

$$\int (x^2 + x^5)\,dx = \int x^2\,dx + \int x^5\,dx = \frac{x^3}{3} + \frac{x^6}{6} + c$$

and $\int (2x^3 + 4)\,dx = 2\int x^3\,dx + \int 4\,dx = 2\left(\frac{x^4}{4}\right) + 4x + c$

$$= \frac{x^4}{2} + 4x + c$$

It should be noted that only one constant c is included since any sum of arbitrary constants gives another arbitrary constant.

Combining rule (ii) with the standard integral for x^n gives:

$$\int ax^n\,dx = \frac{ax^{n+1}}{n+1} + c$$

where a and n are constants and n is **not** equal to -1.

Integrals written in this form are called 'indefinite integrals', since their precise value cannot be found (i.e. c cannot be calculated) unless additional information is provided. (In differentiation there are special rules for multiplication and division of functions. However, there are no such special rules for multiplication and division in integration.)

3 Definite integrals

Limits can be applied to integrals and such integrals are then called 'definite integrals'. The increase in the value of the integral $(x^2 - 3)$ as x increases from 1 to 2 can be written as:

$$\left[\int (x^2 - 3)\,dx\right]_1^2$$

However, this is invariably abbreviated by showing the value of the upper limit at the top of the integral sign and the value of the lower limit at the bottom, i.e.

$$[\int (x^2 - 3)\,dx]_1^2 = \int_1^2 (x^2 - 3)\,dx$$

The integral is evaluated as for an indefinite integral and then placed in the square brackets of the limit operator.

$$\text{Thus} \int_1^2 (x^2 - 3)\,dx = \left[\frac{x^3}{3} - 3x + c\right]_1^2$$

$$= \left[\frac{(2)^3}{3} - 3(2) + c\right] - \left[\frac{(1)^3}{3} - 3(1) + c\right]$$

$$= (\tfrac{8}{3} - 6 + c) - (\tfrac{1}{3} - 3 + c)$$

$$= 2\tfrac{2}{3} - 6 - \tfrac{1}{3} + 3 = -\tfrac{2}{3}$$

The arbitrary constant of integration, c, always cancels out when limits are applied to an integral and it is not usually shown when evaluating a definite integral.

4 Integrals of sin *ax*, cos *ax*, sec² *ax* and e^{ax}

Since integration is the reverse process to that of differentiation the following standard integrals may be deduced.

(a) $\frac{d}{dx}(\sin x) = \cos x$

Hence $\int \cos x\,dx = \sin x + c$

More generally: $\frac{d}{dx}(\sin ax) = a \cos ax$

Hence $\int a \cos ax\,dx = \sin ax + c$

$$\int \cos ax\,dx = \frac{1}{a} \sin ax + c$$

(b) $\frac{d}{dx}(\cos x) = -\sin x$

Hence $\int -\sin x\,dx = \cos x + c$

$$\int \sin x\,dx = -\cos x + c$$

More generally: $\frac{d}{dx}(\cos ax) = -a \sin ax$

Hence $\int -a \sin ax\,dx = \cos ax + c$

$$\int \sin ax\,dx = -\frac{1}{a} \cos ax + c$$

(c) $\frac{d}{dx}(\tan x) = \sec^2 x$

Hence $\int \sec^2 x \, dx = \tan x + c$

More generally: $\frac{d}{dx}(\tan ax) = a \sec^2 ax$

Hence $\int a \sec^2 ax \, dx = \tan ax + c$

$$\int \sec^2 ax \, dx = \frac{1}{a} \tan ax + c$$

(d) $\frac{d}{dx}(e^x) = e^x$

Hence $\int e^x \, dx = e^x + c$

More generally: $\frac{d}{dx}(e^{ax}) = ae^{ax}$

Hence $\int ae^{ax} \, dx = e^{ax} + c$

$$\int e^{ax} \, dx = \frac{1}{a} e^{ax} + c$$

Summary of standard integrals

1. $\int ax^n \, dx = \frac{ax^{n+1}}{n+1} + c$ (except where $n = -1$)
2. $\int \cos ax \, dx = \frac{1}{a} \sin ax + c$
3. $\int \sin ax \, dx = -\frac{1}{a} \cos ax + c$
4. $\int \sec^2 ax \, dx = \frac{1}{a} \tan ax + c$
5. $\int e^{ax} \, dx = \frac{1}{a} e^{ax} + c$
6. $\int \frac{1}{x} \, dx = \ln x + c$

Worked problems on standard integrals

Problem 1. Integrate the following with respect to the variable: (a) x^7; (b) $5.2y^{1.6}$; (c) $\frac{2}{p^3}$.

(a) $\int x^7 \, dx = \frac{x^{7+1}}{7+1} + c = \frac{x^8}{8} + c$

(b) $\int 5.2y^{1.6} \, dy = \frac{5.2y^{1.6+1}}{1.6+1} + c = \frac{5.2y^{2.6}}{2.6} + c = 2.0y^{2.6} + c$

(c) $\displaystyle\int \frac{2}{p^3}\,dp = \int 2p^{-3}\,dp = \frac{2p^{-3+1}}{-3+1} + c$

$$= \frac{2p^{-2}}{-2} + c = \frac{-1}{p^2} + c$$

If the final answer of an integration is differentiated then the original must result (otherwise an error has occurred). For example, in (a) above:

$$\frac{d}{dx}\left(\frac{x^8}{8} + c\right) = \frac{8x^7}{8} = x^7 \text{ (i.e. the original integral)}$$

It will be assumed that in all future integral problems such a check will be made.

Problem 2. Integrate with respect to the variable:
(a) $\left(2x^5 - 4\sqrt{x} + \frac{5}{x^4} - \frac{2}{\sqrt{x^3}} + 6\right)$;
(b) $\left(\frac{4p^5 - 3 + p}{p^3}\right)$.

(a) $\displaystyle\int\left(2x^5 - 4\sqrt{x} + \frac{5}{x^4} - \frac{2}{\sqrt{x^3}} + 6\right)dx = \int (2x^5 - 4x^{\frac{1}{2}} + 5x^{-4} - 2x^{-\frac{3}{2}} + 6)\,dx$

$$= \frac{2x^{5+1}}{(5+1)} - \frac{4x^{\frac{1}{2}+1}}{(\frac{1}{2}+1)} + \frac{5x^{-4+1}}{(-4+1)} - \frac{2x^{-\frac{3}{2}+1}}{(-\frac{3}{2}+1)} + 6x + c$$

$$= \frac{2x^6}{6} - \frac{4x^{\frac{3}{2}}}{\frac{3}{2}} + \frac{5x^{-3}}{-3} - \frac{2x^{-\frac{1}{2}}}{-\frac{1}{2}} + 6x + c$$

$$= \frac{x^6}{3} - \frac{8\sqrt{x^3}}{3} - \frac{5}{3x^3} + \frac{4}{\sqrt{x}} + 6x + c$$

(b) $\displaystyle\int\left(\frac{4p^5 - 3 + p}{p^3}\right)dp = \int\left(\frac{4p^5}{p^3} - \frac{3}{p^3} + \frac{p}{p^3}\right)dp$

$$= \int (4p^2 - 3p^{-3} + p^{-2})\,dp$$

$$= \frac{4p^3}{3} - \frac{3p^{-2}}{-2} + \frac{p^{-1}}{-1} + c$$

$$= \frac{4}{3}p^3 + \frac{3}{2p^2} - \frac{1}{p} + c$$

Problem 3. Given $y = \displaystyle\int\left(r + \frac{1}{r}\right)^2 dr$, find the value of the arbitrary constant of integration if $y = \frac{1}{3}$ when $r = 1$.

$$y = \int\left(r + \frac{1}{r}\right)^2 dr = \int\left(r^2 + 2 + \frac{1}{r^2}\right)dr$$

$$= \frac{r^3}{3} + 2r - \frac{1}{r} + c$$

$y = \frac{1}{3}$ when $r = 1$. Hence $\frac{1}{3} = \frac{(1)^3}{3} + 2(1) - \frac{1}{(1)} + c$

$$\frac{1}{3} = \frac{1}{3} + 2 - 1 + c$$
$$c = -1$$

Hence the arbitrary constant of integration is −1.

Problem 4. Integrate with respect to the variable: (a) $4 \cos 3\theta$; (b) $7 \sin 2x$; (c) $3 \sec^2 5t$.

(a) $\int 4 \cos 3\theta \, d\theta = 4(\frac{1}{3} \sin 3\theta) + c = \mathbf{\frac{4}{3} \sin 3\theta + c}$

(b) $\int 7 \sin 2x \, dx = 7(-\frac{1}{2} \cos 2x) + c = \mathbf{-\frac{7}{2} \cos 2x + c}$

(c) $\int 3 \sec^2 5t \, dt = 3(\frac{1}{5} \tan 5t) + c = \mathbf{\frac{3}{5} \tan 5t + c}$

Problem 5. Find: (a) $\int 6e^{4x} \, dx$; (b) $\int \frac{3}{e^{2t}} \, dt$; (c) $\int \frac{3}{2u} \, du$.

(a) $\int 6e^{4x} \, dx = 6\left(\frac{e^{4x}}{4}\right) + c = \mathbf{\frac{3}{2} e^{4x} + c}$

(b) $\int \frac{3}{e^{2t}} \, dt = \int 3e^{-2t} \, dt = 3\left(\frac{e^{-2t}}{-2}\right) + c = \frac{-3}{2} e^{-2t} + c = \mathbf{\frac{-3}{2e^{2t}} + c}$

(c) $\int \frac{3}{2u} \, du = \frac{3}{2} \int \frac{1}{u} \, du = \mathbf{\frac{3}{2} \ln u + c}$

Problem 6. Evaluate: (a) $\int_1^3 (4x - 3)^2 \, dx$; (b) $\int_0^4 \left(5\sqrt{b} - \frac{1}{\sqrt{b}}\right) db$.

(a) $\int_1^3 (4x - 3)^2 \, dx = \int_1^3 (16x^2 - 24x + 9) \, dx$

$$= \left[\frac{16}{3}x^3 - 24\frac{x^2}{2} + 9x + c\right]_1^3$$

$$= \left[\frac{16}{3}(3)^3 - 12(3)^2 + 9(3) + c\right] - \left[\frac{16}{3}(1)^3 - 12(1)^2 + 9(1) + c\right]$$

$$= (144 - 108 + 27 + c) - (5\tfrac{1}{3} - 12 + 9 + c)$$

$$= (63 + c) - (2\tfrac{1}{3} + c)$$

$$= \mathbf{60\tfrac{2}{3}}$$

The arbitrary constant of integration, c, cancels out, thus showing it to be an unnecessary inclusion when evaluating definite integrals.

(b) $\int_0^4 \left(5\sqrt{b} - \frac{1}{\sqrt{b}}\right) db = \int_0^4 (5b^{\frac{1}{2}} - b^{-\frac{1}{2}}) \, db$

$$= \left[\frac{5b^{\frac{3}{2}}}{\frac{3}{2}} - \frac{b^{\frac{1}{2}}}{\frac{1}{2}}\right]_0^4 = \left[\frac{10}{3}\sqrt{b^3} - 2\sqrt{b}\right]_0^4$$

$$= \left(\frac{10}{3}\sqrt{4^3} - 2\sqrt{4}\right) - \left(\frac{10}{3}\sqrt{0^3} - 2\sqrt{0}\right)$$

$$= \frac{10}{3}(8) - 2(2) - 0 = \frac{80}{3} - 4 = 22\tfrac{2}{3}$$

(taking positive values of square roots only).

Problem 7. Evaluate: (a) $\int_0^{\frac{\pi}{2}} 4 \sin 2x \, dx$; (b) $\int_0^1 3 \cos 3t \, dt$;

(c) $\int_{\frac{\pi}{6}}^{\frac{\pi}{3}} (2 \sin \theta - 3 \cos 2\theta + 4 \sec^2 \theta) \, d\theta$.

(a) $\int_0^{\frac{\pi}{2}} 4 \sin 2x \, dx = \left[-\frac{4}{2} \cos 2x\right]_0^{\frac{\pi}{2}}$

$= \left(-2 \cos 2\left(\frac{\pi}{2}\right)\right) - \left(-2 \cos 2(0)\right)$

$= (-2 \cos \pi) - (-2 \cos 0)$

$= (-2(-1)) - (-2(1))$

$= 2 + 2 = 4$

(b) $\int_0^1 \cos 3t \, dt = [\tfrac{3}{3} \sin 3t]_0^1 = [\sin 3t]_0^1 = (\sin 3 - \sin 0)$

The limits in trigonometric functions are expressed in radians.

Thus 'sin 3' means 'the sine of 3 radians or $3\left(\frac{180}{\pi}\right)^\circ$', i.e. 171.89°.

Hence $\sin 3 - \sin 0 = \sin 171.89^\circ - \sin 0^\circ$

$= 0.141\,1 - 0$

Thus $\int_0^1 3 \cos 3t \, dt = 0.141\,1$

(c) $\int_{\frac{\pi}{6}}^{\frac{\pi}{3}} (2 \sin \theta - 3 \cos 2\theta + 4 \sec^2 \theta) \, d\theta = \left[-2 \cos \theta - \tfrac{3}{2} \sin 2\theta + 4 \tan \theta\right]_{\frac{\pi}{6}}^{\frac{\pi}{3}}$

$= \left(-2 \cos \frac{\pi}{3} - \tfrac{3}{2} \sin \frac{2\pi}{3} + 4 \tan \frac{\pi}{3}\right) - \left(-2 \cos \frac{\pi}{6} - \tfrac{3}{2} \sin \frac{2\pi}{6} + 4 \tan \frac{\pi}{6}\right)$

or $(-2 \cos 60^\circ - \tfrac{3}{2} \sin 120^\circ + 4 \tan 60^\circ) - (-2 \cos 30^\circ - \tfrac{3}{2} \sin 60^\circ + 4 \tan 30^\circ)$

$= (-1 - 1.2990 + 6.9282) - (-1.7321 - 1.2990 + 2.3094)$

$= 5.3509$

Problem 8. Evaluate: (a) $\int_1^2 3e^{4x} \, dx$; (b) $\int_3^4 \frac{5}{x} \, dx$.

(a) $\int_1^2 3e^{4x}\,dx = \left[\frac{3}{4}e^{4x}\right]_1^2 = \frac{3}{4}e^8 - \frac{3}{4}e^4 = \frac{3}{4}e^4(e^4 - 1) = 2\,195$

(b) $\int_3^4 \frac{5}{x}\,dx = 5[\ln x]_3^4 = 5[\ln 4 - \ln 3]$

$= 5 \ln \frac{4}{3} = 1.438\,4$

Further problems on standard integrals may be found in Section 6, Problems 1–65, page 85.

5 Integration by substitution

Functions which require integrating are not usually in the standard integral form previously met. However, by using suitable substitutions some functions can be changed into a form which can be readily integrated. The substitution usually made is to let u be equal to $f(x)$, such that $f(u)\,du$ is a standard integral.

A most important point in the use of substitution is that once a substitution has been made the original variable must be removed completely, because a variable can only be integrated with respect to itself, i.e. we cannot integrate, for example, a function of t with respect to x.

A concept that $\frac{du}{dx}$ is a single entity (measuring the differential coefficient of u with respect to x) has been established in the work done on differentiation. Frequently in work on integration and differential equations, $\frac{du}{dx}$ is split. Provided that when this is done, the original differential coefficient can be re-formed by applying the rules of algebra, then it is in order to do it. For example, if $\frac{dy}{dx} = x$ then it is in order to write $dy = x\,dx$ since dividing both sides by dx re-forms the original differential coefficient. This principle is shown in the following worked problems.

Worked problems on integration by substitution

Problem 1. Find: $\int \cos(5x + 2)\,dx$.

Let $u = 5x + 2$

then $\frac{du}{dx} = 5$, i.e. $dx = \frac{du}{5}$

$\int \cos(5x + 2)\,dx = \int \cos u \frac{du}{5} = \frac{1}{5}\int \cos u\,du$

$= \frac{1}{5}(\sin u) + c$

Since the original integral is given in terms of x, the result should be stated in terms of x.

$u = 5x + 2$

Hence $\int \cos(5x + 2)\,dx = \frac{1}{5}\sin(5x + 2) + c$

Problem 2. Find: $\int (4t - 3)^7 \, dt$.

Let $u = 4t - 3$

then $\frac{du}{dt} = 4$, i.e. $dt = \frac{du}{4}$

$$\int (4t-3)^7 \, dt = \int u^7 \frac{du}{4} = \tfrac{1}{4} \int u^7 \, du$$
$$= \tfrac{1}{4}\left(\frac{u^8}{8}\right) + c$$
$$= \frac{u^8}{32} + c$$

Since $u = (4t - 3)$,

$$\int (4t-3)^7 \, dt = \tfrac{1}{32}(4t-3)^8 + c$$

Problem 3. Integrate $\frac{1}{7x+2}$ with respect to x.

Let $u = 7x + 2$

then $\frac{du}{dx} = 7$, i.e. $dx = \frac{du}{7}$

$$\int \frac{1}{7x+2} \, dx = \int \frac{1}{u}\frac{du}{7}$$
$$= \tfrac{1}{7} \ln u + c$$

Since $u = (7x + 2)$,

$$\int \frac{1}{7x+2} \, dx = \tfrac{1}{7} \ln (7x+2) + c$$

From Problems 1–3 above it may be seen that:

If 'x' in a standard integral is replaced by $(ax + b)$ where a and b are constants, then $(ax + b)$ is written for x in the result and the result is multiplied by $\frac{1}{a}$.

For example, $\int (ax+b) \, dx = \frac{1}{2a}(ax+b)^2 + c$ and, more generally,

$\int (ax+b)^n \, dx = \frac{1}{a(n+1)}(ax+b)^{n+1} + c$ (except when $n = -1$).

Problem 4. Integrate the following with respect to x, using the general rule (i.e. without making a substitution): (a) $3 \sin (2x - 1)$; (b) $2e^{8x+3}$; (c) $\frac{5}{9x-2}$

(a) $\int 3 \sin (2x-1) \, dx = 3(\tfrac{1}{2}) \, [-\cos (2x-1)]$
$= -\tfrac{3}{2} \cos (2x-1) + c$

(b) $\int 2e^{8x+3} \, dx = 2(e^{8x+3})(\tfrac{1}{8}) + c = \tfrac{1}{4} e^{8x+3} + c$

(c) $\int \frac{5}{9x-2}\,dx = 5[\ln(9x-2)]\frac{1}{9} + c = \frac{5}{9}\ln(9x-2) + c$

Problem 5. Find $\frac{3}{2}\int (x^2+2)^6\, 2x\, dx$.

Let $u = x^2 + 2$

then $\frac{du}{dx} = 2x$, i.e. $dx = \frac{du}{2x}$

Hence $\frac{3}{2}\int (x^2+2)^6\, 2x\, dx = \frac{3}{2}\int u^6\, 2x\, \frac{du}{2x} = \frac{3}{2}\int u^6\, du$

The original variable, x, has been removed completely and the integral is now only in terms of u.

$\frac{3}{2}\int u^6\, du = \frac{3}{2}\left(\frac{u^7}{7}\right) + c$

Since $u = x^2 + 2$,

$\int 3x(x^2+2)^6\, dx = \frac{3}{14}(x^2+2)^7 + c$

Problem 6. Find $\int \sin\theta\cos\theta\, d\theta$.

Let $u = \sin\theta$

then $\frac{du}{d\theta} = \cos\theta$, i.e. $d\theta = \frac{du}{\cos\theta}$

Hence $\int \sin\theta\cos\theta\, d\theta = \int u\cos\theta\,\frac{du}{\cos\theta} = \int u\, du = \frac{u^2}{2} + c$

Since $u = \sin\theta$,

$\int \sin\theta\cos\theta\, d\theta = \frac{1}{2}\sin^2\theta + c$

Another solution to this integral is possible.

Let $u = \cos\theta$

then $\frac{du}{d\theta} = -\sin\theta$, i.e. $d\theta = \frac{-du}{\sin\theta}$

Hence $\int \sin\theta\cos\theta\, d\theta = \int \sin\theta\,(u)\left(\frac{-du}{\sin\theta}\right) = -\int u\, du = -\frac{u^2}{2} + c$

Since $u = \cos\theta$,

$\int \sin\theta\cos\theta\, d\theta = -\frac{1}{2}\cos^2\theta + c$

From Problems 5 and 6 above it may be seen that:

Integrals of the form $k\int [f(x)]^n f'(x)\, dx$ (where k is a constant) can be integrated by substituting u for $f(x)$.

Problem 7. Find: $\frac{1}{2}\int \frac{(4x+6)}{\sqrt{(2x^2+6x-1)}}\, dx$

Let $u = 2x^2 + 6x - 1$

then $\frac{du}{dx} = 4x + 6$, i.e. $dx = \frac{du}{4x + 6}$

Hence $\frac{1}{2}\int \frac{(4x + 6)}{\sqrt{(2x^2 + 6x - 1)}}\, dx = \frac{1}{2}\int \frac{(4x + 6)}{\sqrt{u}} \frac{du}{(4x + 6)} = \frac{1}{2}\int \frac{du}{\sqrt{u}}$

$= \frac{1}{2}\int u^{-\frac{1}{2}}\, du = \frac{1}{2}\left(\frac{u^{\frac{1}{2}}}{\frac{1}{2}}\right) + c = u^{\frac{1}{2}} + c$

Since $u = 2x^2 + 6x - 1$,

$$\frac{1}{2}\int \frac{(4x + 6)}{\sqrt{(2x^2 + 6x - 1)}}\, dx = \sqrt{(2x^2 + 6x - 1)} + c$$

Problem 8. Find: $\int \tan\theta\, d\theta$.

$\int \tan\theta\, d\theta = \int \frac{\sin\theta}{\cos\theta}\, d\theta$

Let $u = \cos\theta$

then $\frac{du}{d\theta} = -\sin\theta$, i.e. $d\theta = \frac{-du}{\sin\theta}$

Hence $\int \frac{\sin\theta}{\cos\theta}\, d\theta = \int \frac{\sin\theta}{u}\left(\frac{-du}{\sin\theta}\right) = -\int \frac{1}{u}\, du = -\ln u + c = \ln u^{-1} + c$

Since $u = \cos\theta$,

$\int \tan\theta\, d\theta = \ln(\cos\theta)^{-1} + c$

$= \mathbf{\ln(\sec\theta) + c}$

From Problems 7 and 8 above it may be seen that:

Integrals of the form $k\int \frac{f'(x)}{[f(x)]^n}\, dx$ (where k and n are constants) can be integrated by substituting u for $f(x)$.

Problem 9. Evaluate the following:

(a) $\int_0^1 3\sec^2(4\theta - 1)\, d\theta$

(b) $\int_0^4 5x\sqrt{(2x^2 + 4)}\, dx$, taking positive values of roots only

(c) $\int_1^3 \frac{e^t}{3 + e^t}\, dt$

(a) $\int_0^1 3\sec^2(4\theta - 1)\, d\theta = [\frac{3}{4}\tan(4\theta - 1)]_0^1 = \frac{3}{4}[\tan 3 - \tan(-1)]$

$= \frac{3}{4}[\tan 171.89^\circ - \tan(-57.30^\circ)]$

$= \frac{3}{4}[(-0.142\,5) - (-1.557\,7)]$

$= \frac{3}{4}(1.452\,4) = \mathbf{1.061\,2}$

(b) $\int_0^4 5x\sqrt{(2x^2 + 4)}\,dx = \int_0^4 5x(2x^2 + 4)^{\frac{1}{2}}\,dx$

Let $u = 2x^2 + 4$

then $\frac{du}{dx} = 4x$, i.e. $dx = \frac{du}{4x}$

$\int 5x(2x^2 + 4)^{\frac{1}{2}}\,dx = \int 5x(u^{\frac{1}{2}})\frac{du}{4x} = \frac{5}{4}\int u^{\frac{1}{2}}\,du$

$= \frac{5}{4}\left(\frac{u^{\frac{3}{2}}}{\frac{3}{2}}\right) + c = \frac{5}{6}(\sqrt{u^3}) + c$

Since $u = 2x^2 + 4$,

$\int_0^4 5x\sqrt{(2x^2 + 4)}\,dx = [\frac{5}{6}\sqrt{(2x^2 + 4)^3}]_0^4$

$= \frac{5}{6}\{\sqrt{[(2(4)^2 + 4)]^3} - \sqrt{(4)^3}\}$

$= \frac{5}{6}(216 - 8)$, taking positive values of roots only

$= 173\frac{1}{3}$

(c) $\int_1^3 \frac{e^t}{3 + e^t}\,dt$

Let $u = 3 + e^t$

then $\frac{du}{dt} = e^t$, i.e. $dt = \frac{du}{e^t}$

Hence $\int \frac{e^t}{3 + e^t}\,dt = \int \frac{e^t}{u}\frac{du}{e^t} = \int \frac{du}{u} = \ln u + c$

Since $u = 3 + e^t$,

$\int_1^3 \frac{e^t}{3 + e^t}\,dt = [\ln(3 + e^t)]_1^3$

$= [\ln(3 + e^3) - \ln(3 + e^1)]$

$= \ln\left[\frac{3 + e^3}{3 + e^1}\right] = \ln\left[\frac{23.086}{5.718\,3}\right]$

$= 1.395\,6$

Further problems on integration by substitution may be found in the following Section (6), Problems 66–125, page 89.

6 Further problems

Standard integrals

In Problems 1–35 integrate with respect to the variable.

1. x^5 $\left[\frac{x^6}{6} + c\right]$

2. $2p^3$ $\left[\dfrac{p^4}{2} + c\right]$
3. $3k^6$ $[\frac{3}{7}k^7 + c]$
4. $4u^{2.3}$ $\left[\dfrac{4}{3.3}\,u^{3.3} + c\right]$
5. $x^{-2.1}$ $\left[\dfrac{-x^{-1.1}}{1.1} + c\right]$
6. $\dfrac{2}{x^2}$ $\left[\dfrac{-2}{x} + c\right]$
7. $\dfrac{3}{p}$ $[3 \ln p + c]$
8. $\sqrt{y}$ $[\frac{2}{3}\sqrt{y^3} + c]$
9. $2\sqrt{S^3}$ $[\frac{4}{5}\sqrt{S^5} + c]$
10. $\dfrac{1}{3\sqrt{t}}$ $[\frac{2}{3}\sqrt{t} + c]$
11. $\dfrac{4}{\sqrt[3]{k^2}}$ $[12\sqrt[3]{k} + c]$
12. $3a^3 - \frac{2}{3}\sqrt{a}$ $\left[\dfrac{3a^4}{4} - \frac{4}{9}\sqrt{a^3} + c\right]$
13. $\dfrac{-4}{v^{1.4}}$ $\left[\dfrac{10}{v^{0.4}} + c\right]$
14. $\dfrac{x}{3}(2x + \sqrt{x})$ $\left[\dfrac{2x^3}{9} + \dfrac{2}{15}\sqrt{x^5} + c\right]$
15. $\dfrac{r^3 + 2r - 1}{r^2}$ $\left[\dfrac{r^2}{2} + 2 \ln r + \dfrac{1}{r} + c\right]$
16. $(x + 2)^2$ $\left[\dfrac{x^3}{3} + 2x^2 + 4x + c\right]$
17. $(1 + \sqrt{w})^2$ $\left[w + \frac{4}{3}\sqrt{w^3} + \dfrac{w^2}{2} + c\right]$
18. $\sin 2\theta$ $[-\frac{1}{2}\cos 2\theta + c]$
19. $\cos 4\alpha$ $[\frac{1}{4}\sin 4\alpha + c]$
20. $2 \sin 3t$ $[-\frac{2}{3}\cos 3t + c]$
21. $-4 \cos 5x$ $[-\frac{4}{5}\sin 5x + c]$
22. $\sec^2 6\beta$ $[\frac{1}{6}\tan 6\beta + c]$
23. $-3 \sec^2 t$ $[-3 \tan t + c]$
24. $4(\cos 2\theta - 3 \sin \theta)$ $[2(\sin 2\theta + 6 \cos \theta) + c]$
25. e^{3x} $\left[\dfrac{e^{3x}}{3} + c\right]$
26. $2e^{-4t}$ $[-\frac{1}{2}e^{-4t} + c]$
27. $\dfrac{6}{e^t}$ $\left[-\dfrac{6}{e^t} + c\right]$
28. $3(e^x - e^{-x})$ $[3(e^x + e^{-x}) + c]$
29. $3(e^t - 1)^2$ $\left[3\left(\dfrac{e^{2t}}{2} - 2e^t + t\right) + c\right]$

30. $\dfrac{4}{e^{2x}} + e^x$ $\left[\dfrac{-2}{e^{2x}} + e^x + c\right]$

31. $\dfrac{1}{4t}$ $[\frac{1}{4}\ln t + c]$

32. $\dfrac{3}{5t} + \sqrt{t^5}$ $[\frac{3}{5}\ln t + \frac{2}{7}\sqrt{t^7} + c]$

33. $\left(\dfrac{1}{x} + x\right)^2$ $\left[-\dfrac{1}{x} + 2x + \dfrac{x^3}{3} + c\right]$

34. $3\sin 50\pi t + 4\cos 50\pi t$ $\left[\dfrac{1}{50\pi}(4\sin 50\pi t - 3\cos 50\pi t) + c\right]$

35. $(e^{2x} - 1)(e^{-2x} + 1)$ $[\frac{1}{2}(e^{2x} + e^{-2x}) + c]$

In Problems 36–65 evaluate the definite integrals. (Where roots are involved in the solution, take positive values only when evaluating.)

36. $\displaystyle\int_1^3 2\,dt$ $[4]$

37. $\displaystyle\int_3^5 4x\,dx$ $[32]$

38. $\displaystyle\int_{-4}^2 -3u^2\,du$ $[-72]$

39. $\displaystyle\int_{-1}^1 \tfrac{3}{4}f^2\,df$ $[\frac{1}{2}]$

40. $\displaystyle\int_1^4 x^{-1.5}\,dx$ $[1]$

41. $\displaystyle\int_1^9 \frac{dx}{\sqrt{x}}$ $[4]$

42. $\displaystyle\int_2^5 \frac{4}{x}\,dx$ $[3.665]$

43. $\displaystyle\int_0^2 (x^2 + 2x - 1)\,dx$ $[4\frac{2}{3}]$

44. $\displaystyle\int_1^4 \left(\sqrt{r} - \frac{1}{\sqrt{r}}\right)dr$ $[2\frac{2}{3}]$

45. $\displaystyle\int_1^4 (3x^3 - 4x^2 + x - 2)\,dx$ $[108\frac{3}{4}]$

46. $\displaystyle\int_1^3 (m - 2)(m - 1)\,dm$ $[\frac{2}{3}]$

47. $\displaystyle\int_1^2 \left(\frac{1}{x^2} + \frac{1}{x} + \frac{1}{2}\right)dx$ $[1.693]$

48. $\int_{-2}^{2} (3x - 1)\, dx$ [−4]

49. $\int_{1}^{3} \left(\frac{2}{t^2} - 3t^2 + 4\right) dt$ [$-16\frac{2}{3}$]

50. $\int_{0}^{\frac{\pi}{2}} \sin\theta\, d\theta$ [1]

51. $\int_{0}^{\frac{\pi}{3}} 3\sin 2x\, dx$ [$2\frac{1}{4}$]

52. $\int_{0}^{\frac{\pi}{6}} 4\sin 3\theta\, d\theta$ [$1\frac{1}{3}$]

53. $\int_{\frac{\pi}{6}}^{\frac{\pi}{3}} 2\cos t\, dt$ [0.732 1]

54. $\int_{0}^{1} 5\sin 2\theta\, d\theta$ [3.540 4]

55. $\frac{1}{2}\int_{1}^{2} \cos 3\alpha\, d\alpha$ [−0.070 1]

56. $\int_{0.1}^{0.6} (\frac{1}{4}\sin 3\beta + \frac{1}{2}\cos 2\beta)\, d\beta$ [0.281 9]

57. $\int_{-\frac{\pi}{2}}^{\frac{\pi}{2}} 3\cos\theta\, d\theta$ [6]

58. $\int_{0}^{\frac{\pi}{4}} 3\sec^2\theta\, d\theta$ [3]

59. $\int_{-1}^{1} 3\sec^2 2t\, dt$ [−6.555]

60. $\int_{1}^{2} \frac{e^{3x}}{5}\, dx$ [25.56]

61. $\int_{0.4}^{0.7} 3e^{2t}\, dt$ [2.744]

62. $\int_{0}^{1} \frac{2}{e^{3t}}\, dt$ [0.633 5]

63. $\int_{1}^{4} \left(\frac{t+2}{\sqrt{t}}\right) dt$ [$8\frac{2}{3}$]

64. $\int_{1}^{3} \frac{(3x+2)(x-4)}{x}\, dx$ [−16.789]

65. $\int_{0}^{1} 2\sqrt{x}(x+2)^2\, dx$ [9.105]

In Problems 65–105 integrate with respect to the appropriate variable.

66. $\sin(3x+2)$ $\quad[-\frac{1}{3}\cos(3x+2)+c]$

67. $2\cos(4t+1)$ $\quad[\frac{1}{2}\sin(4t+1)+c]$

68. $3\sec^2(t+5)$ $\quad[3\tan(t+5)+c]$

69. $4\sin(6\theta-3)$ $\quad[-\frac{2}{3}\cos(6\theta-3)+c]$

70. $(2x+1)^5$ $\quad[\frac{1}{12}(2x+1)^6+c]$

71. $3(4S-7)^4$ $\quad[\frac{3}{20}(4S-7)^5+c]$

72. $\frac{1}{12}(9x+5)^8$ $\quad[\frac{1}{972}(9x+5)^9+c]$

73. $\dfrac{1}{3a+1}$ $\quad[\frac{1}{3}\ln(3a+1)+c]$

74. $\dfrac{5}{5f-2}$ $\quad[\ln(5f-2)+c]$

75. $\dfrac{7}{2x+1}$ $\quad[\frac{7}{2}\ln(2x+1)+c]$

76. $\dfrac{-1}{6x+5}$ $\quad[-\frac{1}{6}\ln(6x+5)+c]$

77. $\dfrac{3}{15y-2}$ $\quad[\frac{1}{5}\ln(15y-2)+c]$

78. e^{3x+2} $\quad[\frac{1}{3}e^{3x+2}+c]$

79. $4e^{7t-1}$ $\quad[\frac{4}{7}e^{7t-1}+c]$

80. $2e^{2-3x}$ $\quad[-\frac{2}{3}e^{2-3x}+c]$

81. $4x(2x^2+3)^5$ $\quad[\frac{1}{6}(2x^2+3)^6+c]$

82. $5t(t^2-1)^7$ $\quad[\frac{5}{16}(t^2-1)^8+c]$

83. $(3x^2+4)^8x$ $\quad[\frac{1}{54}(3x^2+4)^9+c]$

84. $\sin^2\theta\cos\theta$ $\quad[\frac{1}{3}\sin^3\theta+c]$

85. $\sin^3 t\cos t$ $\quad[\frac{1}{4}\sin^4 t+c]$

86. $2\cos^2\beta\sin\beta$ $\quad[-\frac{2}{3}\cos^3\beta+c]$

87. $\sec^2\theta\tan\theta$ $\quad[\frac{1}{2}\tan^2\theta+c]$

88. $3\tan 2x\sec^2 2x$ $\quad[\frac{3}{4}\tan^2 2x+c]$

89. $\frac{6}{5}\sin^5\theta\cos\theta$ $\quad[\frac{1}{5}\sin^6\theta+c]$

90. $6x\sqrt{(3x^2+2)}$ $\quad[\frac{2}{3}\sqrt{(3x^2+2)^3}+c]$

91. $(4x^2-1)\sqrt{(4x^3-3x+1)}$ $\quad[\frac{2}{9}\sqrt{(4x^3-3x+1)^3}+c]$

92. $\dfrac{3\ln t}{t}$ $\quad[\frac{3}{2}(\ln t)^2+c]$

93. $\dfrac{6x+2}{(3x^2+2x-1)^5}$ $\quad\left[\dfrac{-1}{4(3x^2+2x-1)^4}+c\right]$

94. $\dfrac{4y-1}{(4y^2-2y+5)^7}$ $\quad\left[\dfrac{-1}{12(4y^2-2y+5)^6}+c\right]$

95. $\dfrac{2x}{\sqrt{(x^2+1)}}$ $\quad[2\sqrt{(x^2+1)}+c]$

96. $\dfrac{3a}{\sqrt{(3a^2+5)}}$ $\quad[\sqrt{(3a^2+5)}+c]$

97. $\dfrac{12x^2+1}{\sqrt{(4x^3+x-1)}}$ $\quad[2\sqrt{(4x^3+x-1)}+c]$

98. $\dfrac{r^2-1}{\sqrt{(r^3-3r+2)}}$ $[\frac{2}{3}\sqrt{(r^3-3r+2)}+c]$

99. $\dfrac{3e^t}{\sqrt{(1+e^t)}}$ $[6\sqrt{(1+e^t)}+c]$

100. $2x\sin(x^2+1)$ $[-\cos(x^2+1)+c]$

101. $(4\theta+1)\sec^2(4\theta^2+2\theta)$ $[\frac{1}{2}\tan(4\theta^2+2\theta)+c]$

102. $\frac{1}{3}(4x+1)\cos(2x^2+x-1)$ $[\frac{1}{3}\sin(2x^2+x-1)+c]$

103. $4te^{2t^2-3}$ $[e^{2t^2-3}+c]$

104. $3\tan\beta$ $[3\ln(\sec\beta)+c]$

105. $(5x-2)e^{5x^2-4x+1}$ $[\frac{1}{2}e^{5x^2-4x+1}+c]$

In Problems 106–125 evaluate the definite integrals.

106. $\displaystyle\int_0^1 (3x-1)^4\,dx$ $[2\frac{1}{5}]$

107. $\displaystyle\int_0^2 (8x-3)(4x^2-3x)^3\,dx$ $[2\,500]$

108. $\displaystyle\int_1^3 x\sqrt{(x^2+1)}\,dx$ $[9.598]$

109. $\displaystyle\int_0^{\frac{\pi}{4}} \sin\left(4\theta+\frac{\pi}{3}\right)d\theta$ $[\frac{1}{4}]$

110. $\displaystyle\int_{\frac{1}{3}}^1 \sec^2(3x-1)\,dx$ $[-0.728\,3]$

111. $\displaystyle\int_1^2 3\cos(5t-2)\,dt$ $[0.508\,9]$

112. $\displaystyle\int_{\frac{1}{2}}^2 \frac{1}{(4s-1)}\,ds$ $[0.486\,5]$

113. $\displaystyle\int_0^2 (9x^2-4)\sqrt{(3x^3-4x)}\,dx$ $[42\frac{2}{3}]$

114. $\displaystyle\int_1^3 \frac{4\ln x}{x}\,dx$ $[2.413\,9]$

115. $\displaystyle\int_0^2 \frac{t}{\sqrt{(2t^2+1)}}\,dt$ $[1]$

116. $\displaystyle\int_1^2 \frac{4x-3}{(2x^2-3x-1)^4}\,dx$ $[-\frac{3}{8}]$

117. $\displaystyle\int_1^2 3\theta\sin(2\theta^2+1)\,d\theta$ $[-0.059\,1]$

118. $\displaystyle\int_0^1 2te^{3t^2-1}\,dt$ $[2.340\,4]$

119. $\displaystyle\int_0^{\frac{\pi}{2}} 3\sin^4\theta\cos\theta\,d\theta$ $\quad[\frac{3}{5}]$

120. $\displaystyle\int_1^2 \frac{dx}{(2x-1)^3}$ $\quad[\frac{2}{9}]$

121. $\displaystyle\int_1^2 \frac{2e^{3\theta}}{e^{3\theta}-5}\,d\theta$ $\quad[2.182\,5]$

122. $\displaystyle\int_0^1 2t\sec^2(3t^2)\,dt$ $\quad[-0.047\,5]$

123. $\displaystyle\int_1^2 x\sin(2x^2-1)\,dx$ $\quad[-0.053\,4]$

124. $\displaystyle\int_{\frac{\pi}{6}}^{\frac{\pi}{3}} \frac{2}{3}\sin t\cos^3 t\,dt$ $\quad[0.083\,3]$

125. $\displaystyle\int_1^2 \frac{e^{3\theta}-e^{-3\theta}}{2}\,d\theta$ $\quad[63.88]$

Chapter 5

Applications of integration to areas, mean values and root mean square values

1 The area between a curve, the x axis and given ordinates

There are several instances in branches of engineering and science where the area under a curve is required to be accurately determined. For example, the areas, between given limits, of:

(a) velocity/time graphs give distances travelled;
(b) force/distance graphs give work done;
(c) acceleration/time graphs give velocities;
(d) voltage/current graphs give power;
(e) pressure/volume graphs give work done;
(f) normal distribution curves give frequencies.

Provided there is a known relationship [e.g. $y = f(x)$] between the variables forming the axes of the above graphs then the areas may be calculated exactly using integral calculus. If a relationship between variables is not known then areas have to be approximately determined using such techniques as the trapezoidal rule, the mid-ordinate rule or Simpson's rule (see Appendix C, page 151).

Let A be the area enclosed between the curve $y = f(x)$, the x axis and the ordinates $x = a$ and $x = b$. Also let A be subdivided into a number of elemental strips each of width δx as shown in Fig. 1.

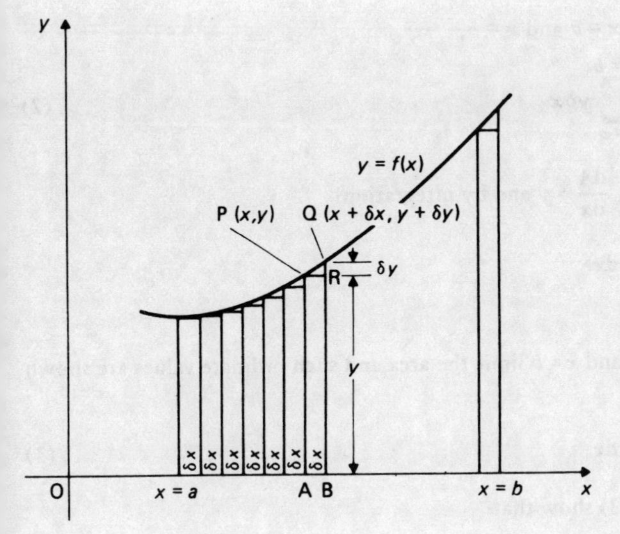

Fig. 1

One such strip is shown as PQRBA, with point P having coordinates (x, y) and point Q having coordinates $(x + \delta x, y + \delta y)$. Let the area PQRBA be δA, which can be seen from Fig. 1 to consist of a rectangle PRBA, of area $y\delta x$, and PQR, which approximates to a triangle of area $\frac{1}{2}\delta x \delta y$,

i.e. $\delta A \simeq y\delta x + \frac{1}{2}\delta x \delta y$

Dividing both sides by δx gives:

$$\frac{\delta A}{\delta x} \simeq y + \tfrac{1}{2}\delta y$$

As δx is made smaller and smaller, the number of rectangles increases and all such areas as PQR become smaller and smaller. Also δy becomes smaller and in the limit as δx approaches zero, $\dfrac{\delta A}{\delta x}$ becomes the differential coefficient $\dfrac{dA}{dx}$ and δy becomes zero,

i.e. $$\lim_{\delta x \to 0}\left(\frac{\delta A}{\delta x}\right) = \frac{dA}{dx} = y + \tfrac{1}{2}(0) = y$$

Hence $$\frac{dA}{dx} = y \qquad \ldots (1)$$

This shows that when a limiting value is taken, all such areas as PQR become zero. Hence the area beneath the curve is given by the sum of all such rectangles as PRBA,

i.e. Area $= \Sigma y \delta x$.

Between the limits $x = a$ and $x = b$,

$$\text{Area, } A = \lim_{\delta x \to 0} \sum_{x=a}^{x=b} y\delta x \qquad \ldots (2)$$

From equation (1), $\dfrac{dA}{dx} = y$ and by integration:

$$\int \frac{dA}{dx}\, dx = \int y\, dx$$

Hence $A = \int y\, dx$

The ordinates $x = a$ and $x = b$ limit the area and such ordinate values are shown as limits.

$$\text{Thus } A = \int_a^b y\, dx \qquad \ldots (3)$$

Equations (2) and (3) show that:

$$\text{Area, } A = \lim_{\delta x \to 0} \sum_{x=a}^{x=b} y\delta x = \int_a^b y\, dx$$

This statement that the limiting value of a sum is equal to the integral between the same limits forms a fundamental theorem of integration. This can be illustrated by considering simple shapes of known areas. For example, Fig. 2 (*a*) shows a rectangle bounded by the line $y = b$, ordinates $x = a$ and $x = b$ and the x-axis.

Let the rectangle be divided into n equal vertical strips of width δx. The area of strip PQAB is $b\delta x$ and since there are n strips making up the total area the total area $= nb\delta x$. The base length of the rectangle, i.e. $(b - a)$, is made up of n strips, each δx in width, hence $n\delta x = (b - a)$. Therefore the total area $= b(b - a)$.

The total area is also obtained by adding the areas of all such strips as PQAB and is independent of the value of n, that is, n can be infinitely large.

$$\text{Hence total area} = \lim_{\delta x \to 0} \sum_{x=a}^{x=b} b\delta x = b(b - a) \qquad \ldots (4)$$

Also the total area is given by $\int_a^b y\, dx = \int_a^b b\, dx$

$$= [bx]_a^b = b(b - a) \qquad \ldots (5)$$

But this is the area obtained from equation (4).

$$\text{Hence } \lim_{\delta x \to 0} \sum_{x=a}^{x=b} b\delta x = \int_a^b b\, dx$$

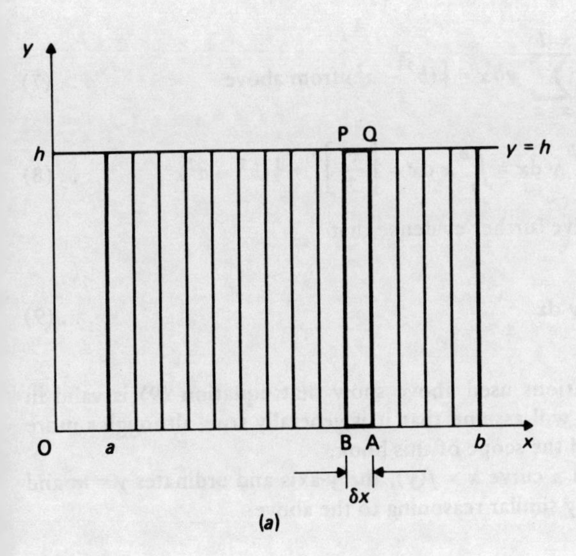

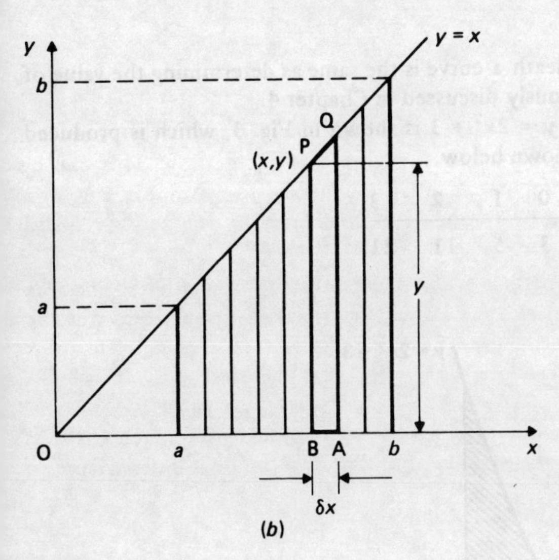

Fig. 2

Similarly for, say, a trapezium bounded by the line $y = x$, the ordinates $x = a$ and $x = b$ and the x-axis (as shown in Fig. 2(b)), the total area is given by:

(half the sum of the parallel sides)(perpendicular distance between these sides)

i.e. $\frac{1}{2}(a + b)(b - a)$ or $\frac{1}{2}(b^2 - a^2)$ . . . (6)

Also, the total area will be given by the sum of all areas such as PQAB which each have an area of $y\delta x$ provided δx is infinitely small.

$$\text{i.e. total area} = \lim_{\delta x \to 0} \sum_{x=a}^{x=b} y\delta x = \tfrac{1}{2}(b^2 - a^2) \text{ from above} \qquad \ldots(7)$$

$$\text{Also, the total area} = \int_a^b y \, dx = \int_a^b x \, dx = \left[\frac{x^2}{2}\right]_a^b = \tfrac{1}{2}(b^2 - a^2) \qquad \ldots(8)$$

Equations (7) and (8) give further evidence that

$$\lim_{\delta x \to 0} \sum_{x=a}^{x=b} y\delta x = \int_a^b y \, dx \qquad \ldots(9)$$

The two simple illustrations used above show that equation (9) is valid in these two cases and we will assume that it is generally true, although a more rigorous proof is beyond the scope of this book.

If the area between a curve $x = f(y)$, the y-axis and ordinates $y = m$ and $y = n$ is required, then by similar reasoning to the above:

$$\text{Area} = \int_m^n x \, dy.$$

Thus finding the area beneath a curve is the same as determining the value of a definite integral as previously discussed in Chapter 4.

A part of the curve $y = 2x^2 + 3$ is shown in Fig. 3, which is produced from the table of values shown below.

x	-2	-1	0	1	2	3
$y = 2x^2 + 3$	11	5	3	5	11	21

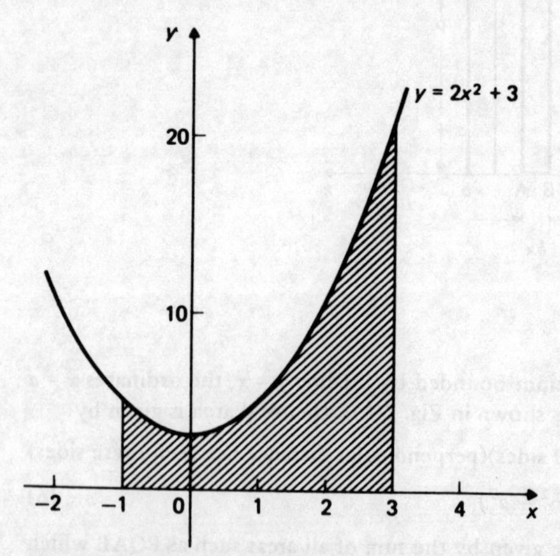

Fig. 3 Graph of $y = 2x^2 + 3$

The area between the curve, the x-axis and the ordinates $x = -1$ and $x = 3$ is shown shaded. This area is given by:

$$\text{Area} = \int_{-1}^{3} y\,dx = \int_{-1}^{3} (2x^2 + 3)\,dx$$

$$= \left[\frac{2x^3}{3} + 3x\right]_{-1}^{3}$$

$$= \left[\frac{2(3)^3}{3} + 3(3)\right] - \left[\frac{2(-1)^3}{3} + 3(-1)\right]$$

$$= \mathbf{30\tfrac{2}{3}\ \textbf{square units}}$$

With the curve $y = 2x^2 + 3$ shown in Fig. 3 all values of y are positive. Hence all the terms in $\Sigma y\delta x$ are positive and $\int_a^b y\,dx$ is positive. However, if a curve should drop below the x-axis, then y becomes negative, all terms in $\Sigma y\delta x$ become negative and $\int_a^b y\,dx$ is negative.

In Fig. 4 the total area between the curve $y = f(x)$, the x-axis and the ordinates $x = a$ and $x = b$ is given by

area P$\left(\text{i.e. } \int_a^c f(x)dx\right)$+ area Q$\left(\text{i.e. } -\int_c^d f(x)dx\right)$+ area R$\left(\text{i.e. } \int_d^b f(x)dx\right)$.

i.e. $\int_a^c f(x)\,dx - \int_c^d f(x)\,dx + \int_d^b f(x)\,dx$

This is **not** the same as the value given by $\int_a^b f(x)\,dx$.

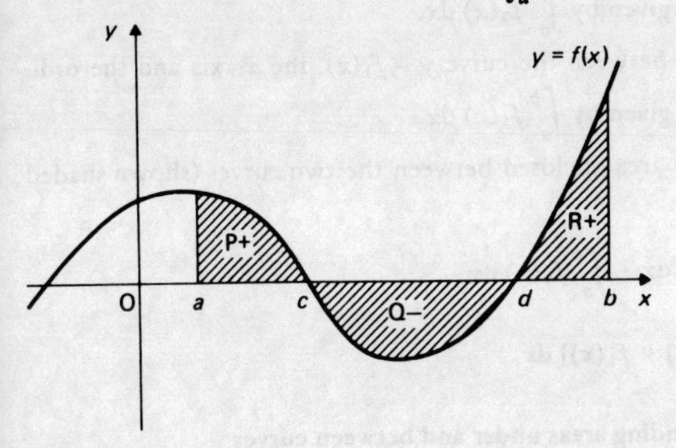

Fig. 4

For this reason, if there is any doubt about the shape of the graph of a function or any possibility of all or part of it lying below the x-axis, a sketch should be made over the required limits to determine if any part of the curve lies below the x-axis.

2 The area between two curves

Let the graphs of the functions $y = f_1(x)$ and $y = f_2(x)$ intersect at points A ($x = a$) and B ($x = b$) as shown in Fig. 5.

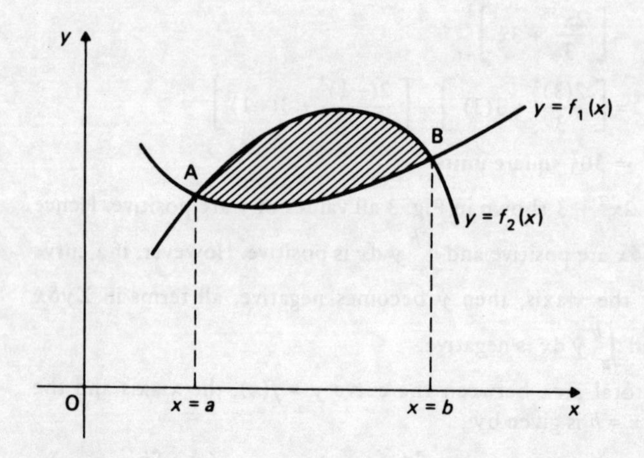

Fig. 5

At the points of intersection $f_1(x) = f_2(x)$.

The area enclosed between the curve $y = f_2(x)$, the x-axis and the ordinates $x = a$ and $x = b$ is given by $\int_a^b f_2(x)\,dx$.

The area enclosed between the curve $y = f_1(x)$, the x-axis and the ordinates $x = a$ and $x = b$ is given by $\int_a^b f_1(x)\,dx$.

It follows that the area enclosed between the two curves (shown shaded in Fig. 5) is given by:

$$\textbf{Shaded area} = \int_a^b f_2(x)\,dx - \int_a^b f_1(x)\,dx$$

$$= \int_a^b [f_2(x) - f_1(x)]\,dx$$

Worked problems on finding areas under and between curves

Problem 1. Sketch the curves and find the areas enclosed by the given curves, the x-axis and the given ordinates: (a) $y = \sin 2x$, $x = 0$, $x = \frac{\pi}{2}$; (b) $y = 3 \cos \frac{1}{2} x$, $x = 0$, $x = \frac{2\pi}{3}$.

(a) A sketch of $y = \sin 2x$ in the range $x = 0$ to $x = \pi$ is shown in Fig. 6(*a*).

The area shown shaded is given by:

$$\text{Area} = \int_0^{\frac{\pi}{2}} \sin 2x \, dx$$

$$= \left[-\frac{\cos 2x}{2} \right]_0^{\frac{\pi}{2}} = \left(\frac{-\cos 2(\pi/2)}{2} \right) - \left(\frac{-\cos 0}{2} \right)$$

$$= \left(\frac{-\cos \pi}{2} \right) - \left(\frac{-\cos 0}{2} \right) = (--\tfrac{1}{2}) - (-\tfrac{1}{2})$$

= 1 square unit

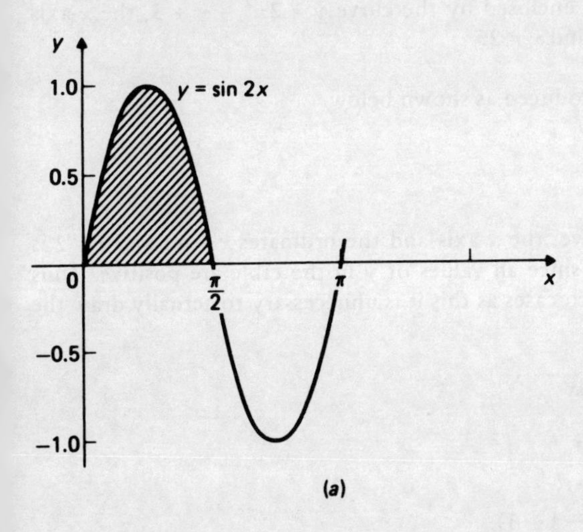

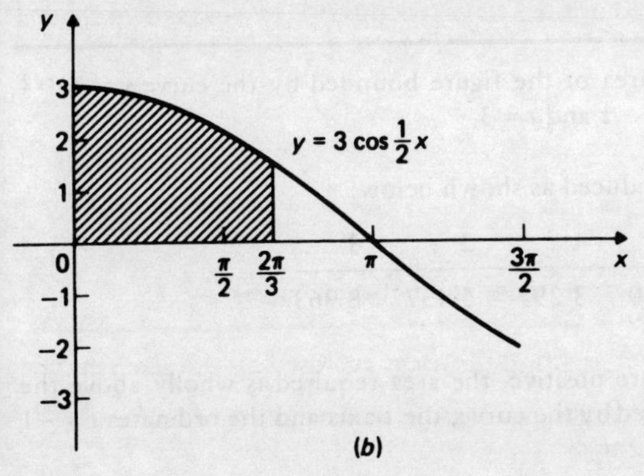

Fig. 6 Graphs of $y = \sin 2x$ and $y = 3 \cos \frac{1}{2}x$

(b) A sketch of $y = 3\cos\frac{1}{2}x$ in the range $x = 0$ to $x = \frac{2\pi}{3}$ is shown in Fig. 6(*b*).

The area shown shaded is given by:

$$\text{Area} = \int_0^{2\pi/3} 3\cos\tfrac{1}{2}x\,dx$$

$$= \left[6\sin\tfrac{1}{2}x\right]_0^{2\pi/3} = \left(6\sin\frac{\pi}{3}\right) - (6\sin 0)$$

$= 6\sin 60° =$ **5.196 square units**

Problem 2. Find the area enclosed by the curve $y = 2x^2 - x + 3$, the x-axis and the ordinates $x = -1$ and $x = 2$.

A table of values is produced as shown below.

x	-1	0	1	2
y	6	3	4	9

The area between the curve, the x-axis and the ordinates $x = -1$ and $x = 2$ is wholly above the x-axis, since all values of y in the table are positive. Thus the area is positive. In such cases as this it is unnecessary to actually draw the graph.

$$\text{Area} = \int_{-1}^{2} (2x^2 - x + 3)\,dx$$

$$= \left[\frac{2x^3}{3} - \frac{x^2}{2} + 3x\right]_{-1}^{2}$$

$$= \left(\tfrac{16}{3} - 2 + 6\right) - \left(-\tfrac{2}{3} - \tfrac{1}{2} - 3\right)$$

$= (9\frac{1}{3}) - (-4\frac{1}{6}) =$ **$13\frac{1}{2}$ square units**

Problem 3. Calculate the area of the figure bounded by the curve $y = 2e^{t/2}$, the t-axis and ordinates $t = -1$ and $t = 3$.

A table of values is produced as shown below.

t	-1	0	1	2	3
$y = 2e^{t/2}$	1.213	2.000	3.297	5.437	8.963

Since all the values of y are positive, the area required is wholly above the t-axis. Hence the area enclosed by the curve, the t-axis and the ordinates $t = -1$ and $t = 3$ is given by:

$$\text{Area} = \int_{-1}^{3} 2e^{t/2}\,dt$$

$$= \left[4e^{t/2}\right]_{-1}^{3} = 4[e^{\frac{3}{2}} - e^{-\frac{1}{2}}]$$

$$= 4[4.481\,7 - 0.606\,5]$$

$$= 15.50 \text{ square units}$$

Problem 4. Find the area enclosed by the curve $y = x^2 + 3$, the x-axis and the ordinates $x = 0$ and $x = 3$. Sketch the curve within these limits. Find also, using integration, the area enclosed by the curve and the y-axis, between the same limits.

A table of values is produced as shown below.

x	0	1	2	3
y	3	4	7	12

(a) Part of the curve $y = x^2 + 3$ is shown in Fig. 7.

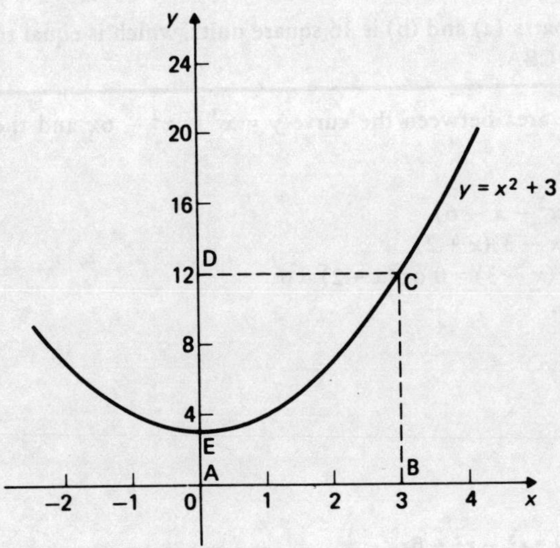

Fig. 7 Graph of $y = x^2 + 3$

The area enclosed by the curve, the x-axis and ordinates $x = 0$ and $x = 3$ (i.e. area ECBA of Fig. 7) is given by:

$$\text{Area} = \int_{0}^{3} (x^2 + 3)\,dx = \left[\frac{x^3}{3} + 3x\right]_{0}^{3}$$

$$= 18 \text{ square units}$$

(b) When $x = 3$, $y = x^2 + 3 = 12$

when $x = 0$, $y = 3$

If $y = x^2 + 3$ then $x^2 = y - 3$ and $x = \sqrt{(y - 3)}$

Hence the area enclosed by the curve $y = x^2 + 3$ (i.e. the curve $x = \sqrt{(y-3)}$), the y-axis and the ordinates $y = 3$ and $y = 12$ (i.e. area EDC of Fig. 7) is given by:

$$\text{Area} = \int_{y=3}^{y=12} x\,dy = \int_3^{12} \sqrt{(y-3)}\,dy$$

Let $u = y - 3$

then $\dfrac{du}{dy} = 1$, i.e. $dy = du$

Hence $\int (y-3)^{\frac{1}{2}}\,dy = \int u^{\frac{1}{2}}\,du = \dfrac{2u^{\frac{3}{2}}}{3}$

Since $u = y - 3$ then

$$\text{Area} = \int_3^{12} \sqrt{(y-3)}\,dy = \left[\tfrac{2}{3}(y-3)^{\frac{3}{2}}\right]_3^{12}$$

$$= \tfrac{2}{3}[\sqrt{9^3} - 0]$$

$$= \textbf{18 square units}$$

The sum of the areas in parts (a) and (b) is 36 square units, which is equal to the area of the rectangle DCBA.

Problem 5. Calculate the area between the curve $y = x^3 - x^2 - 6x$ and the x-axis.

$$y = x^3 - x^2 - 6x = x(x^2 - x - 6)$$
$$= x(x-3)(x+2)$$

Thus when $y = 0$, $x = 0$ or $(x - 3) = 0$ or $(x + 2) = 0$

i.e. $x = 0$, $x = 3$ or $x = -2$.

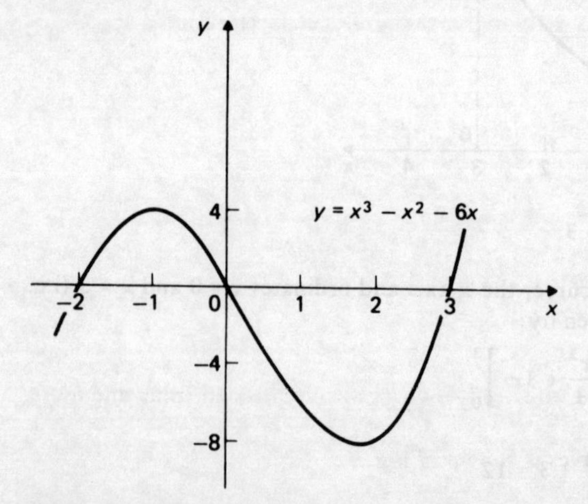

Fig. 8 Graph of $y = x^3 - x^2 - 6x$

Hence the curve cuts the x-axis at $x = 0$, 3 and -2. Since the curve is a continuous function, only one other value need be calculated before a sketch of the curve can be produced. For example, when $x = 1$, $y = -6$, which shows that the portion of the curve between ordinates $x = 0$ and $x = 3$ is negative. Hence the portion of the curve between ordinates $x = 0$ and $x = -2$ must be positive.

A sketch of part of the curve $y = x^3 - x^2 - 6x$ is shown in Fig. 8.
If $y = f(x)$ had not factorised as above, then a table of values could have been produced and the graph sketched in the usual manner.

The sketch shows that the area needs to be calculated in two parts, one part being positive and the other negative, as shown in the second integral below.

The area between the curve and the x-axis is given by:

$$\text{Area} = \int_{-2}^{0} (x^3 - x^2 - 6x)\,dx - \int_{0}^{3} (x^3 - x^2 - 6x)\,dx$$

$$= \left[\frac{x^4}{4} - \frac{x^3}{3} - 3x^2\right]_{-2}^{0} - \left[\frac{x^4}{4} - \frac{x^3}{3} - 3x^2\right]_{0}^{3}$$

$$= (5\tfrac{1}{3}) - (-15\tfrac{3}{4})$$

$$= 21\tfrac{1}{12} \text{ square units}$$

Problem 6. Find the area enclosed between the curves $y = x^2 + 2$ and $y + x = 14$.

The first step is to find the points of intersection of the two curves. This will enable us to limit the range of values when drawing up a table of values in order to sketch the curves. At the points of intersection the curves are equal (i.e. their coordinates are the same). Since $y = x^2 + 2$ and $y + x = 14$ (i.e. $y = 14 - x$) then $x^2 + 2 = 14 - x$ at the points of intersection.

i.e. $x^2 + x - 12 = 0$

$(x - 3)(x + 4) = 0$

Hence $x = 3$ and $x = -4$ at the points of intersection.
Tables of values may now be produced as shown below.

x	-4	-3	-2	-1	0	1	2	3
$y = x^2 + 2$	18	11	6	3	2	3	6	11

x	-4	0	3
$y = 14 - x$	18	14	11

$y = 14 - x$ is a straight line thus only two points are needed (plus one more to check).

A sketch of the two curves is shown in Fig. 9.

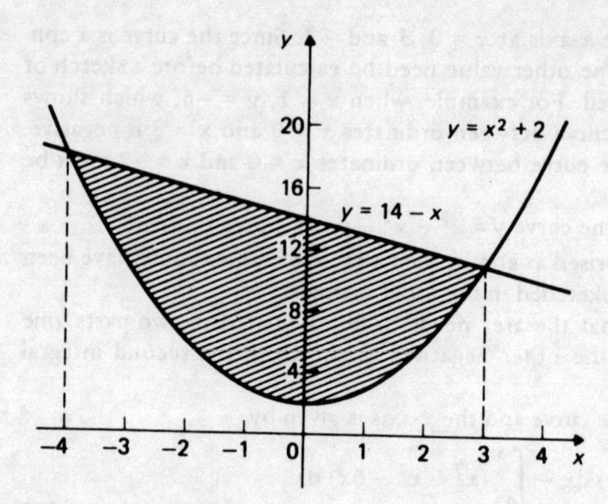

Fig. 9 Graphs of $y = x^2 + 2$ and $y = 14 - x$

The area between the two curves (shown shaded) is given by:

$$\textbf{Shaded area} = \int_{-4}^{3} (14 - x)\,dx - \int_{-4}^{3} (x^2 + 2)\,dx$$

$$= \left[14x - \frac{x^2}{2}\right]_{-4}^{3} - \left[\frac{x^3}{3} + 2x\right]_{-4}^{3}$$

$$= 101\tfrac{1}{2} - 44\tfrac{1}{3}$$

$$= \mathbf{57\tfrac{1}{6}\ square\ units}$$

Problem 7. Find the points of intersection of the two curves $x^2 = 2y$ and $\frac{y^2}{16} = x$. Sketch the two curves and calculate the area enclosed by them.

$x^2 = 2y$, i.e. $y = \frac{x^2}{2}$ or $y^2 = \frac{x^4}{4}$

$\frac{y^2}{16} = x$, i.e. $y^2 = 16x$

At the points of intersection, $\frac{x^4}{4} = 16x$

i.e. $x^4 = 64x$

Hence $x^4 - 64x = 0$

$x(x^3 - 64) = 0$

i.e. $x = 0$ or $x^3 - 64 = 0$

Hence at the points of intersection $x = 0$ and $x = 4$.

Using $y = \frac{x^2}{2}$, when $x = 0$, $y = 0$

when $x = 4$, $y = \frac{(4)^2}{2} = 8$.

[Check, using $y^2 = 16x$. When $x = 0, y = 0$
When $x = 4, y^2 = 64, y = 8$]

Hence the points of intersection of the two curves $x^2 = 2y$ and $\frac{y^2}{16} = x$ are (0, 0) and (4, 8).

A sketch of the two curves (given the special name of parabolas) is shown in Fig. 10.

The area enclosed by the two curves, i.e. OABC (shown shaded), is given by:

$$\text{Area} = \int_0^4 4\sqrt{x}\,dx - \int_0^4 \frac{x^2}{2}\,dx$$

(Note that for one curve $y = \pm 4\sqrt{x}$. The $-4\sqrt{x}$ is neglected since the shaded area required is above the x-axis, and hence positive.)

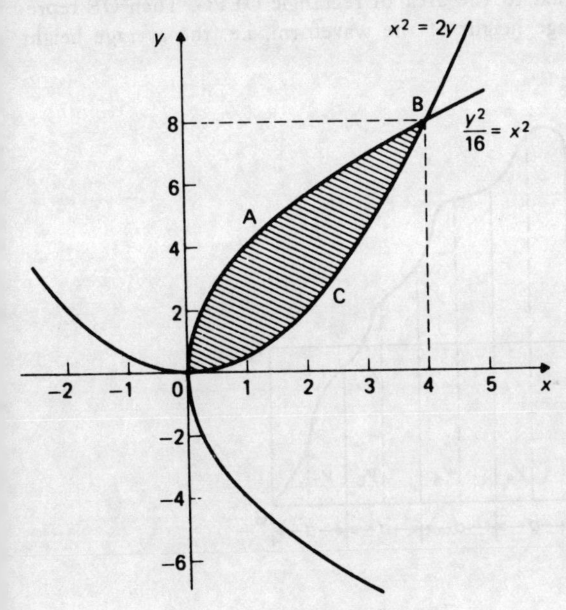

Fig. 10 Graphs of $x^2 = 2y$ and $\frac{y^2}{16} = x$

$$\textbf{Area} = 4\left[\tfrac{2}{3}x^{\frac{3}{2}}\right]_0^4 - \tfrac{1}{2}\left[\frac{x^3}{3}\right]_0^4$$

$$= 21\tfrac{1}{3} - 10\tfrac{2}{3}$$

$$= \mathbf{10\tfrac{2}{3}} \textbf{ square units.}$$

Further problems on areas under and between curves may be found in Section 5, Problems 1 to 48, page 111.

3 Mean or average values

Figure 11 shows the positive half cycle of a periodic waveform of an alternating quantity. If the negative half cycle is the same shape as the positive half cycle then every positive value is balanced by a corresponding negative value and thus the average value of the complete cycle is zero, i.e. the average or mean value over a complete cycle of a symmetrically alternating quantity is zero. However, over half a cycle it has a non zero value.

Let the area of the waveform in Fig. 11 representing the positive half cycle be divided into, say, 7 strips each of width d, with ordinates at the mid-point of each strip (mid-ordinates) represented by y_1, y_2, y_3 and so on. Let EF be drawn parallel to base OG such that the area under the curve between O and G is equal to the area of rectangle OEFG. Then OE represents the mean or average height of the waveform, i.e. the average height of the y ordinates.

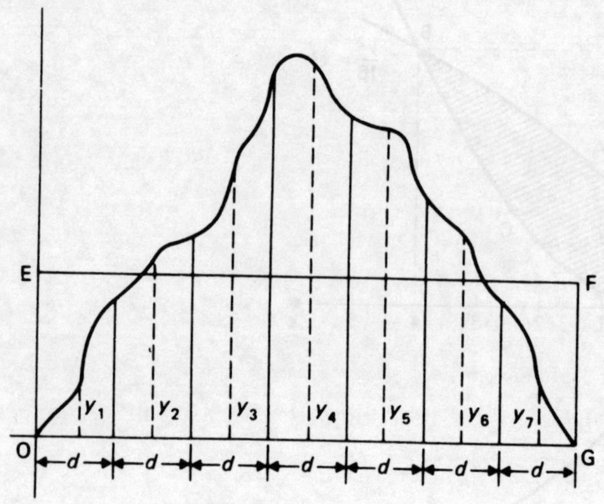

Fig. 11

From the mid-ordinate rule (see Appendix C, page 151):

$$\text{Area under curve} = d\,(y_1 + y_2 + y_3 + y_4 + y_5 + y_6 + y_7).$$

Also, area of rectangle OEFG = (OE)(OG)

= (average value) ($7d$)

But area under curve = area of rectangle OEFG.

Thus $d\,(y_1 + y_2 + y_3 + y_4 + y_5 + y_6 + y_7) = (\text{average value})\,(7d)$

Let the average value be denoted by $\bar{y}$ (pronounced y bar).

$$\text{Then } \bar{y} = \frac{d\,(y_1 + y_2 + y_3 + y_4 + y_5 + y_6 + y_7)}{7d}$$

$$\text{i.e.}\quad \bar{y} = \frac{\textbf{area under curve}}{\textbf{length of base}}$$

In the example shown in Fig. 11 the mid-ordinate rule is used to find the area under the curve, although other approximate methods such as Simpson's rule or the trapezoidal rule (see Appendix C) could equally well have been used.

An exact method of finding areas under curves is that of integration discussed in Section 1, although this is only possible if (a) there is an equation relating the variables, and (b) the equation can be integrated.

Figure 12 shows part of a curve $y = f(x)$. The mean value, $\overline{y}$, of the curve between the limits $x = a$ and $x = b$ is given by:

$$\overline{y} = \frac{\text{area under curve}}{\text{length of base}} = \frac{\text{area PSRQ}}{b - a}$$

From Section 1, area under the curve $y = f(x)$ between the limits $x = a$ and $x = b$ is given by:

$$\int_a^b f(x)\,\mathrm{d}x$$

$$\text{Hence } \overline{y} = \frac{\int_a^b f(x)\,\mathrm{d}x}{b - a} = \frac{1}{b - a}\int_a^b y\,\mathrm{d}x$$

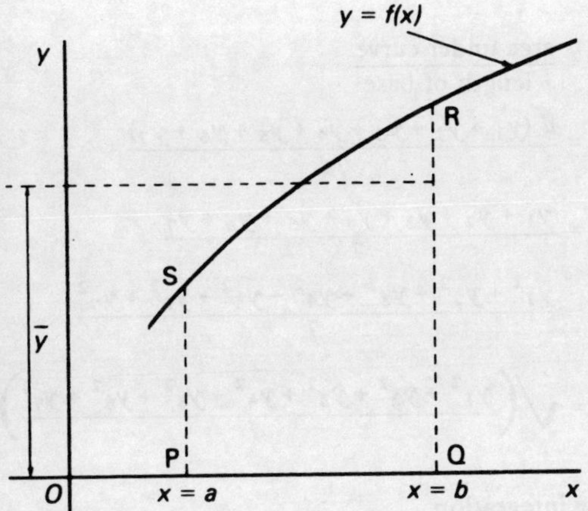

y = f(x)
y
R
S
$\overline{y}$
P
Q
O
x = a
x = b
x

Fig. 12

4 Root mean square values

The root mean square value of a quantity is 'the square root of the average value of the squared values of the quantity' taken over an interval. In many scientific applications – particularly those involving periodic waveforms – mean values, when determined, are found to be zero because there are equal numbers of positive and negative values which cancel each other out. In such cases the root mean square (r.m.s.) values can be valuable e.g.:

(a) the average rate of the heating effect of an electric current (i.e. proportional to current^2),
(b) the standard deviation, used in statistics to estimate the spread or scatter of a set of data (i.e. proportional to distance^2), and
(c) the average linear velocity of a particle in a body which is rotating about an axis (i.e. proportional to velocity^2).

Each of these applications depend upon **square values**, which do not cancel. For example, a direct current I amperes passing through a resistor R ohms for t seconds produces a heating effect given by I^2Rt joules. When an alternating current i amperes is passed through the same resistor R for the same time, the instantaneous value of the heating effect, i.e. $i^2\,Rt$, varies. In order to give the same heating effect as the equivalent direct current, i^2 is replaced by the **mean square value.** Then the r.m.s. value of an alternating current is defined as that current which will give the same heating effect as the equivalent direct current, I amperes.

The r.m.s. value obtained without using integration

Referring to Fig. 11.

Average or mean value, $\bar{y}$ $= \dfrac{\text{area under curve}}{\text{length of base}}$

$$= \frac{d\,(y_1 + y_2 + y_3 + y_4 + y_5 + y_6 + y_7)}{7d}$$

$$= \frac{y_1 + y_2 + y_3 + y_4 + y_5 + y_6 + y_7}{7}$$

Average value of the squares of the function $= \dfrac{y_1^2 + y_2^2 + y_3^2 + y_4^2 + y_5^2 + y_6^2 + y_7^2}{7}$

The square root of the average value of the squares of the function, i.e. the r.m.s. value $= \sqrt{\left(\dfrac{y_1^2 + y_2^2 + y_3^2 + y_4^2 + y_5^2 + y_6^2 + y_7^2}{7}\right)}$

The r.m.s. value obtained using integration

The average value $\bar{y} = \dfrac{1}{b-a}\displaystyle\int_a^b y\,\mathrm{d}x$ (from section 3)

The average value of the square of the function $= \dfrac{1}{b-a}\displaystyle\int_a^b y^2\,\mathrm{d}x$

The square root of the average value of the squares of the function, i.e. the r.m.s. value $= \sqrt{\left[\dfrac{1}{b-a}\displaystyle\int_a^b y^2\,\mathrm{d}x\right]}$

One of the principle applications of r.m.s. values is in alternating currents and voltages in electrical engineering. Alternating waveforms are frequently of the form $i = I_m \sin\theta$ or $v = V_m \sin\theta$, and, when determining r.m.s. values, integrals of the form $\int (I_m \sin\theta)^2 \, d\theta$ result, i.e. it is necessary to be able to integrate $\sin^2\theta$ and also $\cos^2\theta$.

From Appendix B, section 3, equation 12 (page 151):

$$\cos 2A = 1 - 2\sin^2 A$$

Thus $$\sin^2 A = \frac{1 - \cos 2A}{2}$$

Hence $$\int \sin^2 A \, dA = \int \frac{1 - \cos 2A}{2} \, dA = \frac{1}{2}\left[A - \frac{\sin 2A}{2}\right] + c$$

From Appendix B, section 3, equation 11 (page 151)

$$\cos 2A = 2\cos^2 A - 1$$

Thus $$\cos^2 A = \frac{1 + \cos 2A}{2}$$

Hence $$\int \cos^2 A \, dA = \int \frac{1 + \cos 2A}{2} \, dA = \frac{1}{2}\left[A + \frac{\sin 2A}{2}\right] + c$$

Worked problems on mean and r.m.s. values

Problem 1. Determine (a) the mean value and (b) the r.m.s. value of $y = 3x^2$ between $x = 1$ and $x = 3$, using integration.

(a) Mean value, $$\bar{y} = \frac{1}{3-1}\int_1^3 3x^2 \, dx$$

$$= \frac{1}{2}\left[x^3\right]_1^3 = \frac{1}{2}(27 - 1) = \mathbf{13}$$

(b) r.m.s. value $$= \sqrt{\left[\frac{1}{3-1}\int_1^3 (3x^2)^2 \, dx\right]}$$

$$= \sqrt{\left[\frac{1}{2}\int_1^3 9x^4 \, dx\right]}$$

$$= \sqrt{\left\{\frac{9}{2}\left[\frac{x^5}{5}\right]_1^3\right\}}$$

$$= \sqrt{\left\{\frac{9}{10}[243 - 1]\right\}} = \sqrt{\left[\frac{9\,(242)}{10}\right]}$$

$$= \sqrt{217.8} = \mathbf{14.76}$$

Problem 2. A sinusoidal electrical current is given by $i = 10.0 \sin\theta$ amperes. Determine the mean value of the current over half a cycle using integration.

Average or mean value, $$\bar{y} = \frac{1}{\pi - 0}\int_0^\pi 10.0 \sin\theta \, d\theta$$

$$= \frac{10.0}{\pi}\left[-\cos\theta\right]_0^\pi$$

$$= \frac{10.0}{\pi}\,[(-\cos\pi) - (-\cos 0)]$$

$$= \frac{10.0}{\pi}\,[(--1) - (-1)] = \frac{10.0}{\pi}\,(2)$$

$$= \frac{2}{\pi} \times 10.0 = \mathbf{6.366\ amperes.}$$

Note that for a sine wave, the mean value $= \dfrac{2}{\pi} \times$ maximum value.

Problem 3. Using the current given in problem 2, determine the r.m.s. value using integration.

$$\text{r.m.s. value} = \sqrt{\left\{\frac{1}{\pi}\int_0^{\pi}(10.0\sin\theta)^2\,d\theta\right\}}$$

$$= \sqrt{\left\{\frac{100.0}{\pi}\int_0^{\pi}\sin^2\theta\,d\theta\right\}}$$

From Appendix B, section 3, $\cos 2\theta = 1 - 2\sin^2\theta$

$$\text{from which } \sin^2\theta = \frac{1-\cos 2\theta}{2}$$

$$\text{Hence r.m.s. value} = \sqrt{\left\{\frac{100.0}{\pi}\int_0^{\pi}\frac{1-\cos 2\theta}{2}\,d\theta\right\}}$$

$$= \sqrt{\left\{\left(\frac{100.0}{\pi}\right)\frac{1}{2}\left[\theta - \frac{\sin 2\theta}{2}\right]_0^{\pi}\right\}}$$

$$= \sqrt{\left\{\left(\frac{100.0}{\pi}\right)\frac{1}{2}\left[\left(\pi - \frac{\sin 2\pi}{2}\right) - \left(0 - \frac{\sin 2(0)}{2}\right)\right]\right\}}$$

$$= \sqrt{\left\{\frac{100.0}{\pi}\left(\frac{\pi}{2}\right)\right\}} = \frac{10.0}{\sqrt{2}} = \mathbf{7.071\ amperes.}$$

Note that for a sine wave, the r.m.s. value = $\dfrac{1}{\sqrt{2}} \times$ maximum value.

Problem 4. Find the area bounded by the curve $y = 6x - x^2$ and the x axis for values of x from 0 to 6. Determine also the mean value and the r.m.s. value of y over the same range.

A table of values is drawn up as shown below.

$y = 6x - x^2$

x	0	1	2	3	4	5	6
y	0	5	8	9	8	5	0

Since all the values of y are positive the area required is wholly above the x axis. Hence the area enclosed by the curve, the x axis and the ordinates $x = 0$ and $x = 6$ is given by:

$$\text{Area} = \int_0^6 y\,dx = \int_0^6 (6x - x^2)\,dx$$

$$= \left[\frac{6x^2}{2} - \frac{x^3}{3}\right]_0^6 = (108 - 72) - (0 - 0) = \mathbf{36\ square\ units.}$$

$$\text{Mean or average value} = \frac{\text{area under curve}}{\text{length of base}} = \frac{36}{6} = \mathbf{6}$$

$$\text{r.m.s. value} = \sqrt{\left\{\frac{1}{6-0}\int_0^6 (6x - x^2)^2\,dx\right\}}$$

$$= \sqrt{\left\{\frac{1}{6}\int_0^6 (36x^2 - 12x^3 + x^4)\,dx\right\}}$$

$$= \sqrt{\left\{\frac{1}{6}\left[\frac{36x^3}{3} - \frac{12x^4}{4} + \frac{x^5}{5}\right]_0^6\right\}}$$

$$= \sqrt{\left\{\frac{1}{6}\left[\left(12(6)^3 - 3(6)^4 + \frac{(6)^5}{5}\right) - (0)\right]\right\}}$$

$$= \sqrt{\left\{\frac{1}{6}(2\,592 - 3\,888 + 1\,555.2)\right\}}$$

$$= \sqrt{\left\{\frac{1}{6}(259.2)\right\}} = \sqrt{43.2} = \mathbf{6.573}$$

Further problems on mean and r.m.s. values may be found in the following section (5), problems 49 to 83, page 114.

5 Further problems

Areas under and between curves.

All answers are in square units.

In problems 1 to 22 find the area enclosed between the given curve, the horizontal axis and the given ordinates. Sketch the curve in the given range for each.

1. $y = 2x; x = 0, x = 5$ [25]
2. $y = x^2 - x + 2; x = -1, x = 2$ [$7\frac{1}{2}$]
3. $y = \frac{x}{2}; x = 3, x = 7$ [10]
4. $y = p - 1; p = 1, p = 5$ [8]
5. $F = 8S - 2S^2; S = 0, S = 2$ [$10\frac{2}{3}$]
6. $y = (x - 1)(x - 2); x = 0, x = 3$ [$1\frac{5}{6}$]
7. $y = 8 + 2x - x^2; x = -2, x = 4$ [36]
8. $u = 2(4 - t^2); t = -2, t = 2$ [$21\frac{1}{3}$]
9. $y = x(x - 1)(x + 3); x = -2, x = 1$ [$7\frac{11}{12}$]
10. $x = 4a^3; a = -2, a = 2$ [32]
11. $y = x(x - 1)(x - 3); x = 0, x = 3$ [$3\frac{1}{12}$]

12. $a = t^3 + t^2 - 4t - 4$; $t = -3$, $t = 3$ $\quad [24\frac{1}{3}]$
13. $y = \sin\theta$; $\theta = 0$, $\theta = \frac{\pi}{2}$ $\quad [1]$
14. $y = \cos x$; $x = \frac{\pi}{4}$, $x = \frac{\pi}{2}$ $\quad [0.292\ 9]$
15. $y = 3\sin 2\beta$; $\beta = 0$, $\beta = \frac{\pi}{4}$ $\quad [1\frac{1}{2}]$
16. $y = 5\cos 3\alpha$; $\alpha = 0$, $\alpha = \frac{\pi}{6}$ $\quad [1\frac{2}{3}]$
17. $y = \sin x - \cos x$; $x = 0$, $x = \frac{\pi}{4}$ $\quad [0.414\ 2]$
18. $2y^2 = x$; $x = 0$, $x = 2$ $\quad [2\frac{2}{3}]$
19. $5 = xy$; $x = 2$, $x = 5$ $\quad [4.581]$
20. $y = 2e^{2t}$; $t = 0$, $t = 2$ $\quad [53.60]$
21. $ye^{4x} = 3$; $x = 1$, $x = 3$ $\quad [0.013\ 7]$
22. $y = 2x + e^x$; $x = 0$, $x = 3$ $\quad [28.09]$
23. Find the area between the curve $y = 3x - x^2$ and the x axis. $[4\frac{1}{2}]$
24. Calculate the area enclosed between the curve $y = 12 - x - x^2$ and the x axis using integration. $[57\frac{1}{6}]$
25. Sketch the curve $y = \sec^2 2x$ from $x = 0$ to $x = \frac{\pi}{4}$ and calculate the area enclosed between the curve, the x axis and the ordinates $x = 0$ and $x = \frac{\pi}{6}$. $[0.866]$
26. Find the area of the template enclosed between the curve $y = \frac{1}{x-2}$, the x axis and the ordinates $x = 3$ cm and $x = 5$ cm. $[1.098\ 6\ \text{cm}^2]$
27. Sketch the curves $y = x^2 + 4$ and $y + x = 10$ and find the area enclosed by them. $[20\frac{5}{6}]$
28. Calculate the area enclosed between the curves $y = \sin\theta$ and $y = \cos\theta$ and the θ axis between the limits $\theta = 0$ and $\theta = \frac{\pi}{4}$. $[0.414\ 2]$
29. Find the area between the two parabolas $9y^2 = 16x$ and $x^2 = 6y$. $[3\frac{5}{9}]$
30. Calculate the area of the metal plate enclosed between $y = x\ (x - 4)$ and the x axis where x is in metres. $[10\frac{2}{3}\ \text{m}^2]$
31. Sketch the curve $x^2 - y = 3x + 10$ and find the area enclosed between it and the x axis. $[57\frac{1}{6}]$
32. Find the area enclosed by the curve $y = 4\ (x^2 - 1)$, the x axis and the ordinates $x = 0$ and $x = 2$. Find also the area enclosed by the curve and the y axis between the same limits. $[8, 21\frac{1}{3}]$
33. Calculate the area between the curve $y = x\ (x^2 - 2x - 3)$ and the x axis using integration. $[11\frac{5}{6}]$
34. Find the area of the figure bounded by the curve $y = 3e^{2x}$, the x axis and the ordinates $x = -2$ and $x = 2$. $[81.87]$
35. Find the area enclosed between the curves $y = x^2 - 3x + 5$ and $y - 1 = 2x$. $[4\frac{1}{2}]$

36. Find the points of intersection of the two curves $\frac{x^2}{2} = \sqrt{2y}$ and $y^2 = 8x$ and calculate the area enclosed by them. [(0,0), (4,5.657); 7.542]

37. Calculate the area bounded by the curve $y = x^2 + x + 4$ and the line $y = 2(x + 5)$. [$20\frac{5}{6}$]

38. Find the area enclosed between the curves $y = x^2$ and $y = 8 - x^2$. [$21\frac{1}{3}$]

39. Calculate the area between the curve $y = 3x^3$ and the line $\frac{y}{12} = x$ in the first quadrant. [12]

40. Find the area bounded by the three straight lines $y = 4(2 - x)$, $y = 4x$ and $3y = 4x$. [2]

41. A vehicle has an acceleration a of $(30 + 2t)$ metres per second after t seconds. If the vehicle starts from rest find its velocity after 10 seconds.

(Velocity $= \int_{t_1}^{t_2} a \, dt$) [400 m s^{-1}]

42. A car has a velocity v of $(3 + 4t)$ metres per second after t seconds. How far does it move in the first 4 seconds? Find the distance travelled in the fifth second.

(Distance travelled $= \int_{t_1}^{t_2} v \, dt$) [44 m; 21 m]

43. A gas expands according to the law pv = constant. When the volume is 2 m^3 the pressure is 200 kPa. Find the work done as the gas expands from 2 m^3 to a volume of 5 m^3.

(Work done $= \int_{v_1}^{v_2} p \, dv$) [367 kJ]

44. The brakes are applied to a train and the velocity v at any time t seconds after applying the brakes is given by $(16 - 2.5\,t)$ m s^{-1}. Calculate the distance travelled in 8 seconds.

(Distance travelled $= \int_{t_1}^{t_2} v \, dt$) [48 m]

45. The force F newtons acting on a body at a distance x metres from a fixed point is given by $F = 3x + \frac{1}{x^2}$. Find the work done when the body moves from the position where $x = 1$ m to that where $x = 3$ m.

(Work done $= \int_{x_1}^{x_2} F \, dx$) [$12\frac{2}{3}$ newton metres]

46. The velocity v of a body t seconds after a certain instant is $(4t^2 + 3)$ m s^{-1} Find how far it moves in the interval from $t = 2$ s to $t = 6$ s.

(Distance travelled $= \int_{t_1}^{t_2} v \, dt$) [$289\frac{1}{3}$ m]

47. The heat required to raise the temperature of carbon dioxide from 300 K to 600 K is determined from the area formed when the heat capacity

(C_p) is plotted against the temperature (T) between 300 K and 600 K. If $C_p = 27 + 42 \times 10^{-3}\ T - 14.22 \times 10^{-6}\ T^2$, determine the area by integration. [12 870]

48. The entropy required to raise hydrogen sulphide from 400 K to 500 K is determined from the area formed when C_p is plotted against the temperature (T) between 400 K and 500 K. Given that $C_p = 37 + 0.008\ T$, determine the area by integration. [4060]

Mean and r.m.s. values.

In problems 49 to 55 find the mean values over the ranges stated.

49. $y = 2\sqrt{x}$ from $x = 0$ to $x = 4$. $[2\frac{2}{3}]$
50. $y = t\,(2 - t)$ from $t = 0$ to $t = 2$. $[\frac{2}{3}]$
51. $y = \sin\theta$ from $\theta = 0$ to $\theta = 2\pi$. [0]
52. $y = \sin\theta$ from $\theta = 0$ to $\theta = \pi$. $\left[\frac{2}{\pi} \text{ or } 0.637\right]$
53. $y = 2\cos 2x$ from $x = 0$ to $x = \frac{\pi}{4}$. $\left[\frac{4}{\pi} \text{ or } 1.273\right]$
54. $y = 2e^x$ from $x = 1$ to $x = 4$. [34.59]
55. $y = \frac{2}{x}$ from $x = 1$ to $x = 3$. [1.099]
56. Determine the mean value of the curve $y = t - t^2 + 2$ which lies above the t axis by the mid-ordinate rule and check your result using integration. $[1\frac{1}{2}]$
57. The velocity v of a piston moving with simple harmonic motion at any time t is given by $v = k \sin \omega t$. Find the mean velocity between $t = 0$ and $t = \frac{\pi}{\omega}$. $\left[\frac{2k}{\pi}\right]$
58. Calculate the mean value of $y = 3x - x^2$ in the range $x = 0$ to $x = 3$ by integration. [1.5]
59. If the speed v m s^{-1} of a car is given by $v = 3\,t + 5$, where t is the time in seconds, find the mean value of the speed from $t = 2$ s to $t = 5$ s. $[15\frac{1}{2}$ m s$^{-1}]$
60. The number of atoms N remaining in a mass of material during radioactive decay after time t seconds is given by $N = N_o\, e^{-\lambda t}$, where N_o and λ are constants. Determine the mean number of atoms in the mass of material for the time period $t = 0$ to $t = \frac{1}{\lambda}$. $[0.632\ N_o]$
61. A force $9\sqrt{x}$ newtons acts on a body whilst it moves from $x = 0$ to $x = 4$ metres. Find the mean value of the force with respect to distance x. [12 N]
62. The rotor of an electric motor has a tangential velocity v given by $v = (9 - t^2)$ metres per second after t seconds. Find how far a point on the circumference of the rotor moves in 3 seconds from $t = 0$ and the average velocity during this time. [18 m, 6 m/s]
63. The vertical height y kilometres of a rocket fired from a launcher varies

with the horizontal distance x kilometres and is given by $y = 6x - x^2$. Determine the mean height of the rocket from $x = 0$ to $x = 6$ kilometres. [6 km]

In problems 64 to 71 find the r.m.s. values over the ranges stated.

64. $y = 2x$ from $x = 0$ to $x = 4$. [4.619]

65. $y = x^2$ from $x = 1$ to $x = 3$. [4.919]

66. $y = \sin t$ from $t = 0$ to $t = 2\pi$. $\left[\frac{1}{\sqrt{2}} \text{ or } 0.707\right]$

67. $y = \sin t$ from $t = 0$ to $t = \pi$. $\left[\frac{1}{\sqrt{2}} \text{ or } 0.707\right]$

68. $y = 4 + 2\cos x$ from $x = 0$ to $x = 2\pi$. [4.243]

69. $y = \sin 3\theta$ from $\theta = 0$ to $\theta = \frac{\pi}{6}$. $\left[\frac{1}{\sqrt{2}} \text{ or } 0.707\right]$

70. $y = 1 + \sin t$ from $t = 0$ to $t = 2\pi$. [1.225]

71. $y = \cos\theta - \sin\theta$ from $\theta = 0$ to $\theta = \frac{\pi}{4}$. [0.603]

72. Determine: (a) the average value; and (b) the r.m.s. value of a sine wave of maximum value 5.0 for:
(i) a half cycle; and
(ii) one cycle

(a) (i) [3.18] (ii) [0]
(b) (i) [3.54] (ii) [3.54]

73. The distances of points, y, from the mean value of a frequency distribution are related to the variate, x, by the equation $y = x + \frac{1}{x}$. Determine the standard deviation (i.e. the r.m.s. value), correct to 4 significant figures, for values of x from 1 to 2. [2.198]

74. Show that the ratio of the r.m.s. value to the mean value of $y = \sin x$ over the period $x = 0$ to $x = \pi$ is given by $\frac{\pi}{2\sqrt{2}}$.

75. Draw the graph of $4t - t^2$ for values of t from 0 to 4. Determine the area bounded by the curve and the t axis by integration. Find also the mean and r.m.s. values over the same range.
[Area = $10\frac{2}{3}$ square units; mean value = 2.67; r.m.s. value = 2.92]

76. An alternating voltage is given by $v = 20.0 \cos 50\pi t$ volts. Find: (a) the mean value; and (b) the r.m.s. value over the interval from $t = 0$ to $t = 0.01$ seconds. [12.73 V; 14.14 V]

77. A voltage, $v = 24 \sin 50\pi t$ volts is applied across an electrical circuit. Find its mean and r.m.s. values over the range $t = 0$ to $t = 10$ ms, each correct to 4 significant figures. [15.28 V, 16.97 V]

78. A sinusoidal voltage has a maximum value of 150 volts. Calculate its r.m.s. and mean values. [106.1 V, 95.49 V]

79. In a frequency distribution the average distance from the mean, p, is related to the variable, q, by the equation $p = 3q^2 - 2$. Determine the r.m.s. deviation from the mean for values of q from -2 to $+3$, correct to 3 significant figures. [8.66]

80. If the dipolar coupling (y) between two parallel magnetic dipoles in a liquid is given by $y = 1 - 3x^2$ determine the average value of y between $x = 1$ and -1. [0]

81. Determine the average heat capacity $\overline{c}_p$ of magnesium between 300 K and 400 K given that:

$$c_p = 6.2 + 1.3 \times 10^{-3}\ T - 6.8 \times 10^4\ \mathrm{T}^{-2}$$

[6.09]

82. If the rate of a chemical reaction (r) is given by:

$$r = 2.5\ (3.2 - x)\ (3 - x),$$

where x is the moles of the product, determine the average rate for x to increase from 0 to 1 mole. [17.1]

83. Find the average velocity ($\overline{v}$) of a chemical change during the first 5 minutes of reaction when $v = e^{-3t}$, where t is the time in minutes. $\left[\dfrac{1 - e^{-15}}{15}\right]$

Chapter 6

Numerical integration

1 Introduction

In Chapter 4 it was shown that a number of functions, such as ax^n, $\sin ax$, $\cos ax$, e^{ax} and $\frac{1}{x}$, are termed '**standard integrals**', and they may be integrated 'on sight'. It was also shown that with some other simple functions, such as $\cos(5x + 2)$, $(4t - 3)^7$ and $\frac{1}{7x + 2}$, an **algebraic substitution** may be used to change the function into a form which can be readily integrated. However, even with more advanced methods of integration, there are many mathematical functions which cannot be integrated by analytical methods and thus approximate methods have then to be used. Also, in some cases, only a set of observed tabulated numerical values of the function to be integrated may be available. Approximate values of definite integrals may be determined by what is termed **numerical integration.** It is shown in Chapters 4 and 5 that determining the value of a definite integral is, in fact, finding the area between a curve, the horizontal axis and the specified ordinates. Three methods of finding approximate areas under curves are the trapezoidal rule, the mid-ordinate rule and Simpson's rule and these rules are used as a basis for numerical integration.

2 The trapezoidal rule

Let a required definite integral be denoted by $\int_a^b y\,dx$ and be represented

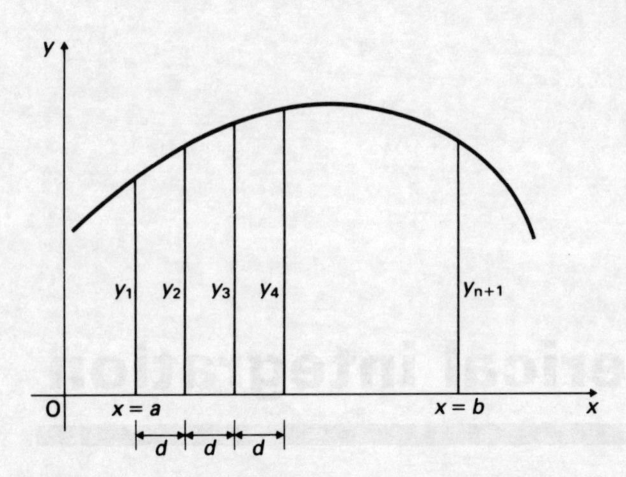

Fig. 1

by the area under the graph of y between the limits $x = a$ and $x = b$, as shown in Fig. 1. Let the range of integration be divided into n equal intervals each of width d, such that $n\text{d} = b - a$, i.e., $\text{d} = \dfrac{b-a}{n}$.

Let the ordinates be labelled $y_1, y_2, \ldots y_{n+1}$, as shown. An approximation to the area under the curve is obtained by joining the tops of the ordinates by straight lines. Each strip of area is thus a trapezium and since the area of a trapezium is given by:

area of trapezium = $\frac{1}{2}$ (sum of the parallel sides) (perpendicular distance between them)

then

$$\int_a^b y\,dx \approx \frac{1}{2}(y_1 + y_2)\text{d} + \frac{1}{2}(y_2 + y_3)\text{d} + \frac{1}{2}(y_3 + y_4)\text{d} + \ldots + \frac{1}{2}(y_n + y_{n+1})\text{d}$$

$$\approx \text{d}\left[\frac{1}{2}y_1 + y_2 + y_3 + \ldots + y_n + \frac{1}{2}y_{n+1}\right]$$

$$\text{i.e., } \int_a^b y\,dx \approx \begin{pmatrix}\text{width of}\\ \text{interval}\end{pmatrix} \left\{\frac{1}{2}\begin{pmatrix}\text{first + last}\\ \text{ordinate}\end{pmatrix} + \text{sum of remaining ordinates}\right\} \ldots \quad (1)$$

To demonstrate this method of numerical integration let an integral be chosen whose value may be determined exactly using integration.

Evaluating $\int_1^3 \frac{1}{\sqrt{x}}\,dx$ by integration gives $\left[\frac{x^{-\frac{1}{2}+1}}{-\frac{1}{2}+1}\right]_1^3 = \left[2\sqrt{x}\right]_1^3$

$= 2(\sqrt{3} - \sqrt{1})$

= **1.464**, correct to 3 decimal places.

The range of integration is the difference between the upper and lower limits, i.e., 3 − 1 = 2. Using the trapezoidal rule with, say, 4 intervals, gives an interval width, d of $\frac{3-1}{4} = \frac{1}{2}$, and ordinates situated at 1.0, 1.5, 2.0, 2.5 and 3.0. Corresponding values of $\frac{1}{\sqrt{x}}$ are as shown in the table below, each given correct to 4 decimal places.

x	1.0	1.5	2.0	2.5	3.0
$\frac{1}{\sqrt{x}}$	1.0000	0.8165	0.7071	0.6325	0.5774

From equation (1):

$$\int_1^3 \frac{1}{\sqrt{x}}\,dx \approx \left(\frac{1}{2}\right)\left[\frac{1}{2}(1.0000 + 0.5774) + 0.8165 + 0.7071 + 0.6325\right]$$

= **1.472**, correct to 3 decimal places.

Using the trapezoidal rule with, say, 8 intervals, each of width $\frac{3-1}{8}$, i.e., 0.25, gives ordinates situated at 1.00, 1.25, 1.50, 1.75, 2.00, 2.25, 2.50, 2.75 and 3.00. Corresponding values of $\frac{1}{\sqrt{x}}$ are as shown in the table below.

x	1.00	1.25	1.50	1.75	2.00	2.25	2.50	2.75	3.00
$\frac{1}{\sqrt{x}}$	1.0000	0.8944	0.8165	0.7559	0.7071	0.6667	0.6325	0.6030	0.5774

From equation (1):

$$\int_1^3 \frac{1}{\sqrt{x}}\,dx \approx (0.25)\left[\frac{1}{2}(1.0000 + 0.5774) + 0.8944 + 0.8165 + 0.7559 + 0.7071 + 0.6667 + 0.6325 + 0.6030\right]$$

= **1.466**, correct to 3 decimal places.

 The greater the number of intervals chosen (i.e., the smaller the interval width) the more accurate will be the value of the definite integral. The exact value is found when the number of intervals is infinite, i.e., when the interval width d tends to zero, and this is of course what the process of integration is based upon.

3 The mid-ordinate rule

Let a required definite integral be denoted again by $\int_a^b y \, dx$ and represented by the area under the graph of y between the limits $x = a$ and $x = b$, as shown in Fig. 2.

With this rule each strip of width d is assumed to be replaced by a rectangle of height equal to the ordinates at the middle point of each interval, shown as $y_1, y_2, y_3, \ldots y_n$ in Fig. 2. Thus,

$$\int_a^b y \, dx \approx d\,y_1 + d\,y_2 + d\,y_3 + \ldots + d\,y_n$$

$$\approx d\,(y_1 + y_2 + y_3 + \ldots + y_n)$$

i.e., $$\int_a^b y \, dx \approx \textbf{(width of interval) (sum of mid-ordinates)} \qquad \ldots (2)$$

The more intervals chosen the more accurate will be the value of the definite integral.

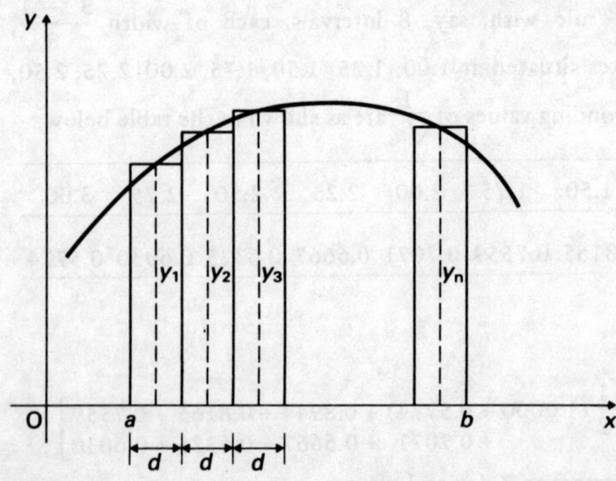

Fig. 2

Applying this rule to evaluating $\int_1^3 \frac{1}{\sqrt{x}}\,dx$ with, say, 4 intervals means that the width interval d is $\frac{3-1}{4}$, i.e., $\frac{1}{2}$ and that ordinates exist at 1.0, 1.5, 2.0, 2.5 and 3.0. Hence mid-ordinates y_1, y_2, y_3 and y_4 occur at 1.25, 1.75, 2.25 and 2.75. Corresponding values of $\frac{1}{\sqrt{x}}$ are shown in the following table:

x	1.25	1.75	2.25	2.75
$\frac{1}{\sqrt{x}}$	0.8944	0.7559	0.6667	0.6030

From equation (2):

$$\int_1^3 \frac{1}{\sqrt{x}}\,dx \approx (\tfrac{1}{2})\ [0.8944 + 0.7559 + 0.6667 + 0.6030]$$

$$= \mathbf{1.460}, \quad \text{correct to 3 decimal places.}$$

Using the mid-ordinate rule with, say, 8 intervals, each of width 0.25, gives ordinates at 1.00, 1.25, 1.50, 1.75 . . . and thus mid-ordinates at 1.125, 1.375, 1.625, 1.875, ... Corresponding values of $\frac{1}{\sqrt{x}}$ are shown in the following table.

x	1.125	1.375	1.625	1.875	2.125	2.375	2.625	2.875
$\frac{1}{\sqrt{x}}$	0.9428	0.8528	0.7845	0.7303	0.6860	0.6489	0.6172	0.5898

From equation (2);

$$\int_1^3 \frac{1}{\sqrt{x}}\,dx \approx (0.25)\begin{bmatrix} 0.9428 + 0.8528 + 0.7845 + 0.7303 + 0.6860 \\ + 0.6489 + 0.6172 + 0.5898 \end{bmatrix}$$

$$= \mathbf{1.463}, \text{ correct to 4 decimal places.}$$

As before, the greater the number of values chosen the nearer the result will be to the true one.

4 Simpson's rule

In section 2, it is shown that the approximation made with the trapezoidal rule is to join the tops of two successive ordinates by a straight line, i.e., by using a linear approximation of the form $a + bx$. With Simpson's rule, the approximation made is to join the tops of three successive ordinates by a parabola, i.e., by using a quadratic approximation of the form $a + bx + cx^2$.

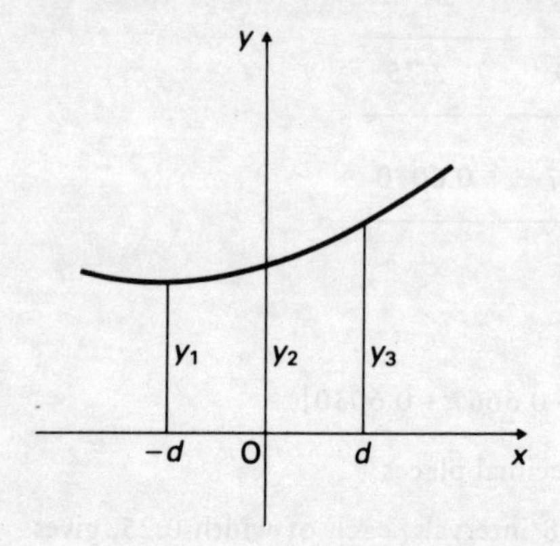

Fig. 3

Figure 3 shows three ordinates, y_1, y_2 and y_3 of a parabola $y = a + bx + cx^2$ at x = -d, x = 0 and x = d respectively. Thus the width of each of the two intervals is d. The area under the parabola from x = −d to x = d is given by:

$$\int_{-d}^{d} (a + bx + cx^2)\,dx = \left[ax + \frac{bx^2}{2} + \frac{cx^3}{3}\right]_{-d}^{d}$$

$$= (ad + \frac{bd^2}{2} + \frac{cd^3}{3}) - (-ad + \frac{bd^2}{2} - \frac{cd^3}{3})$$

$$= 2\,ad + \frac{2}{3}\,cd^3 \quad = \frac{1}{3}\,d\,(6a + 2\,cd^2) \qquad \ldots (3)$$

Since $y = a + bx + cx^2$, at $x = -d$, $y_1 = a - bd + cd^2$,

at $x = 0$, $y_2 = a$,

and at $x = d$, $y_3 = a + bd + cd^2$.

Hence $y_1 + y_3 = 2a + 2cd^2$

and $y_1 + 4y_2 + y_3 = 6a + 2cd^2 \qquad \ldots (4)$

Thus the area under the parabola between $x = -d$ and $x = d$ in Fig. 3 is (from equations (3) and (4)):

$$\frac{1}{3} d (y_1 + 4y_2 + y_3).$$

and this result can be seen to be independent of the position of the origin.

Let a definite integral be denoted by $\int_a^b y \, dx$ and represented by the area under the graph of y between the limits $x = a$ and $x = b$, as shown in Fig. 4. The range of integration, $b - a$, is divided into an even number of intervals, say, $2n$, each of width d.

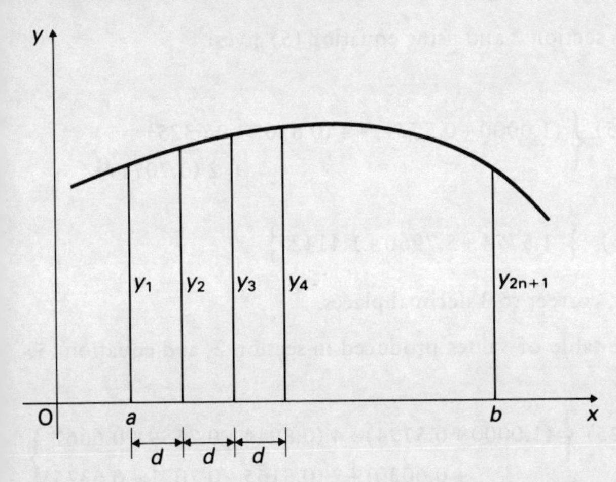

Fig. 4

Since an even number of intervals is specified, an odd number of ordinates, $2n + 1$, exists. Let an approximation to the curve over the first two intervals be a parabola of the form $y = a + bx + cx^2$ which passes through the tops of the three ordinates y_1, y_2 and y_3. Similarly, let an approximation to the curve over the next two intervals be the parabola which passes through the tops of the ordinates y_3, y_4 and y_5, and so on.

$$\text{Then } \int_a^b y \, dx \approx \frac{1}{3} d (y_1 + 4y_2 + y_3) + \frac{1}{3} d (y_3 + 4y_4 + y_5) + \ldots + \frac{1}{3} d (y_{2n-1} + 4y_{2n} + y_{2n+1})$$

$$\approx \frac{1}{3} d \left[(y_1 + y_{2n+1}) + 4 (y_2 + y_4 + \ldots + y_{2n}) + 2(y_3 + y_5 + \ldots + y_{2n-1}) \right]$$

i.e., $$\int_a^b y\,dx \approx \frac{1}{3}\begin{pmatrix}\text{width of}\\ \text{interval}\end{pmatrix}\left\{\begin{pmatrix}\text{first + last}\\ \text{ordinate}\end{pmatrix} + 4\begin{pmatrix}\text{sum of even}\\ \text{ordinates}\end{pmatrix} + 2\begin{pmatrix}\text{sum of remaining}\\ \text{odd ordinates}\end{pmatrix}\right\} \quad \ldots (5)$$

Again, the more intervals chosen the more accurate will be the value of the definite integral.

Applying this rule to evaluate $\int_1^3 \frac{1}{\sqrt{x}}\,dx$ with, say, 4 intervals, using the table of values produced in section 2 and using equation (5) gives:

$$\int_1^3 \frac{1}{\sqrt{x}}\,dx \approx \frac{1}{3}(0.5)\left\{(1.0000 + 0.5774) + 4\,(0.8165 + 0.6325) + 2\,(0.7071)\right\}$$

$$\approx \frac{1}{3}(0.5)\left\{1.5774 + 5.7960 + 1.4142\right\}$$

$= \mathbf{1.465}$, correct to 3 decimal places.

Using 8 intervals, the table of values produced in section 2, and equation (5) gives:

$$\int_1^3 \frac{1}{\sqrt{x}}\,dx \approx \frac{1}{3}(0.25)\left\{(1.0000 + 0.5774) + 4\,(0.8944 + 0.7559 + 0.6667 + 0.6030) + 2\,(0.8165 + 0.7071 + 0.6325)\right\}$$

$$\approx \frac{1}{3}(0.25)\left\{1.5774 + 11.6800 + 4.3122\right\}$$

$= \mathbf{1.464}$, correct to 3 decimal places, which is the same value as obtained by integration.

Simpson's rule is generally regarded as the most accurate of the three approximate methods used in numerical integration.

Worked problems on numerical integration

Problem 1. Evaluate $\int_0^1 \frac{2}{1 + x^2}\,dx$, using the trapezoidal rule with 8 intervals, giving the answer correct to 3 decimal places.

The range of integration is from 0 to 1 and hence if 8 intervals are chosen then each will be of width $\frac{1-0}{8}$ i.e., d = 0.125. Thus ordinates occur at 0, 0.125, 0.250, 0.375 . . . and for each of these values $\frac{2}{1+x^2}$ may be evaluated. The results are shown in the following table.

x	0	0.125	0.250	0.375	0.500	0.625	0.750	0.875	1.000
$\frac{2}{1+x^2}$	2.0000	1.9692	1.8824	1.7534	1.6000	1.4382	1.2800	1.1327	1.0000

(Note that since 3-decimal place accuracy is required in the final answer, values in the table are taken correct to one decimal place more than this, i.e., correct to 4 decimal places.)

From equation (1), using the trapezoidal rule with 8 intervals:

$$\int_0^1 \frac{2}{1+x^2}\, dx \approx d\left[\frac{1}{2}\begin{pmatrix}\text{first + last}\\ \text{ordinate}\end{pmatrix} + \text{sum of remaining ordinates}\right]$$

$$\approx (0.125)\left[\frac{1}{2}(2.0000 + 1.0000) + 1.9692 + 1.8824 + 1.7534 + 1.6000 + 1.4382 + 1.2800 + 1.1327\right]$$

$$= \mathbf{1.569}, \text{ correct to 3 decimal places.}$$

Problem 2. Determine the value of $\int_1^5 \ln x\, dx$ using the mid-ordinate rule with (a) 4 intervals, (b) 8 intervals. Give the answers correct to 4 significant figures.

(a) The range of integration is from 1 to 5 and hence if 4 intervals are chosen then each will be of width $\frac{5-1}{4}$, i.e., d = 1.

Hence ordinates occur at 1, 2, 3, 4 and 5 and thus mid-ordinates occur at 1.5, 2.5, 3.5 and 4.5. Corresponding values of ln x are shown in the table below.

x	1.5	2.5	3.5	4.5
ln x	0.40547	0.91629	1.25276	1.50408

From equation (2), using the mid-ordinate rule with 4 intervals:

$$\int_1^5 \ln x \, dx \approx \text{(width of interval) (sum of mid-ordinates)}$$
$$\approx (1)(0.40547 + 0.91629 + 1.25276 + 1.50408)$$
$$= \mathbf{4.079}, \text{ correct to 4 significant figures.}$$

(b) With 8 intervals, width $d = \dfrac{5-1}{8} = 0.5$, ordinates occur at 1.0, 1.5, 2.0, 2.5, . . . 5, and thus mid-ordinates occur at 1.25, 1.75, 2.25, . . . 4.75

Hence from equation (2):

$$\int_1^5 \ln x \, dx \approx (0.5)\,[\ln 1.25 + \ln 1.75 + \ln 2.25 + \ln 2.75 + \ln 3.25 + \ln 3.75 + \ln 4.25 + \ln 4.75]$$
$$= \mathbf{4.055}, \text{ correct to 4 significant figures.}$$

Problem 3. Evaluate $\displaystyle\int_0^{\frac{\pi}{2}} \frac{1}{1 + \frac{1}{2}\sin^2\theta}\, d\theta$, using Simpson's rule with 6 intervals, correct to 3 decimal places.

With 6 intervals chosen, each will have a width of $\dfrac{\frac{\pi}{2} - 0}{6}$, i.e. $\dfrac{\pi}{12}$ rad (or 15°) and the ordinates occur at 0, $\dfrac{\pi}{12}, \dfrac{\pi}{6}, \dfrac{\pi}{4}, \dfrac{\pi}{3}, \dfrac{5\pi}{12}$ and $\dfrac{\pi}{2}$.

Corresponding values of $\dfrac{1}{1 + \frac{1}{2}\sin^2\theta}$ are evaluated and are shown in the table below.

θ	0	$\frac{\pi}{12}$ (or 15°)	$\frac{\pi}{6}$ (or 30°)	$\frac{\pi}{4}$ (or 45°)	$\frac{\pi}{3}$ (or 60°)	$\frac{5\pi}{12}$ (or 75°)	$\frac{\pi}{2}$ (or 90°)
$\dfrac{1}{1+\frac{1}{2}\sin^2\theta}$	1.0000	0.9676	0.8889	0.8000	0.7273	0.6819	0.6667

From equation (5), using Simpson's rule with 6 intervals:

$$\int_0^{\frac{\pi}{2}} \frac{1}{1 + \frac{1}{2}\sin^2\theta}\, d\theta \approx \frac{1}{3}\begin{pmatrix}\text{width of}\\ \text{intervals}\end{pmatrix}\left\{\begin{pmatrix}\text{first + last}\\ \text{ordinate}\end{pmatrix} + 4\begin{pmatrix}\text{sum of even}\\ \text{ordinates}\end{pmatrix} + 2\begin{pmatrix}\text{sum of remaining}\\ \text{odd ordinates}\end{pmatrix}\right\}$$

$$\approx \frac{1}{3}\left(\frac{\pi}{12}\right)\left\{(1.0000+0.6667)+4\,(0.9676+0.8000+0.6819)+2\,(0.8889+0.7273)\right\}$$

$$\approx \frac{\pi}{36}\left\{1.6667 + 9.7980 + 3.2324\right\}$$

$$\approx 0.4083\pi = \mathbf{1.283}, \text{ correct to 3 decimal places.}$$

Problem 4. The velocity v of a car has the following values for corresponding values of time t from $t = 0$ to $t = 8$ s.

v m s^{-1}	0	0.6	1.7	2.8	4.9	7.0	9.2	10.8	12.0
t s	0	1	2	3	4	5	6	7	8

The distance travelled by the car in 8 s is given by $\int_0^8 v\,dt$.

Determine the approximate distance travelled by using (a) the trapezoidal rule, and (b) Simpson's rule, using 8 intervals in each case.

Since 8 intervals are chosen, each has a width of 1 s.

(a) From equation (1), using the trapezoidal rule with 8 intervals:

$$\int_0^8 v\,dt \approx (1)\left[\frac{1}{2}(0 + 12.0) + 0.6 + 1.7 + 2.8 + 4.9 + 7.0 + 9.2 + 10.8\right]$$

$$= \mathbf{43\ m}$$

(b) From equation (5), using Simpson's rule with 8 intervals:

$$\int_0^8 v\,dt \approx \frac{1}{3}(1)\left\{(0+12.0)+4\,(0.6+2.8+7.0+10.8)+2\,(1.7+4.9+9.2)\right\}$$

$$\approx \frac{1}{3}(1)\ [12.0+84.8+31.6]$$

$$= \mathbf{42.8\ m}$$

Problem 5. Evaluate $\int_0^{1.2} e^{-\frac{x^2}{2}}\,dx$ using (a) the trapezoidal rule, (b) the mid-ordinate rule and (c) Simpson's rule. Use 6 intervals in each case and give answers correct to 3 significant figures.

Since 6 intervals are chosen then each is of width $\frac{1.2 - 0}{6}$, i.e., 0.2

Hence ordinates occur at 0, 0.2, 0.4, 0.6, 0.8, 1.0 and 1.2

Corresponding values of $e^{-\frac{x^2}{2}}$ are evaluated and shown in the following table.

x	0	0.2	0.4	0.6	0.8	1.0	1.2
$e^{-\frac{x^2}{2}}$	1.0000	0.9802	0.9231	0.8353	0.7261	0.6065	0.4868

(a) From equation (1), using the trapezoidal rule with 6 intervals:

$$\int_0^{1.2} e^{-\frac{x^2}{2}}\,dx \approx (0.2)\left[\frac{1}{2}(1.0000+0.4868)+0.9802+0.9231 +0.8353+0.7261+0.6065\right]$$

$$= \mathbf{0.963}, \text{ correct to 3 significant figures.}$$

(b) Mid-ordinates occur at 0.1, 0.3, 0.5, 0.7, 0.9 and 1.1 and corresponding values of $e^{-\frac{x^2}{2}}$ are evaluated and shown in the following table.

x	0.1	0.3	0.5	0.7	0.9	1.1
$e^{-\frac{x}{2}}$	0.9950	0.9560	0.8825	0.7827	0.6670	0.5461

From equation (2), using the mid-ordinate rule with 6 intervals:

$$\int_0^{2} e^{-\frac{x^2}{2}}\,dx \approx (0.2)(0.9950+0.9560+0.8825+0.7827+0.6670 +0.5461)$$

$$= \mathbf{0.966}, \text{ correct to 3 significant figures.}$$

(c) From equation (5), using Simpson's rule with 6 intervals and the table of values in part (a) above:

$$\int^{1.2} e^{-\frac{x^2}{2}}\,dx \approx \frac{1}{3}(0.2)\left\{(1.0000+0.4868)+4\,(0.9802+\ 8353 +0.6065)+2\,(0.9231+0.7261)\right\}$$

$$\approx \frac{1}{3}(0.2)\,[1.4868+9.6880+3.2984]$$

$$= \mathbf{0.965}, \text{ correct to 3 significant figures.}$$

Problem 6. Evaluate, correct to 3 decimal places, $\int_1^3 \frac{5}{x}\,dx$ using

(a) integration,
(b) the trapezoidal rule, with (i) 4 intervals, (ii) 8 intervals,
(c) Simpson's rule with (i) 4 intervals, (ii) 8 intervals.
(d) Determine the percentage error in parts (b) and (c) compared with the value obtained in part (a).

(a) $\int_1^3 \frac{5}{x}\,dx = 5[\ln x]_1^3 = 5[\ln 3 - \ln 1] = 5 \ln 3 = \mathbf{5.493}$, correct to 3 decimal places.

(b) (i) With 4 intervals, each is of width $\frac{3-1}{4} = 0.5$ and ordinates occur at 1.0, 1.5, 2.0, 2.5 and 3.0. Corresponding values of $\frac{5}{x}$ are evaluated and shown in the following table.

x	1.0	1.5	2.0	2.5	3.0
$\frac{5}{x}$	5.0000	3.3333	2.5000	2.0000	1.6667

From equation (1), using the trapezoidal rule with 4 intervals:

$$\int_1^3 \frac{5}{x}\,dx \approx (0.5)\left[\frac{1}{2}(5.00000 + 1.6667) + 3.3333 + 2.5000 + 2.0000\right]$$
$$= \mathbf{5.583}, \text{ correct to 3 decimal places.}$$

(ii) With 8 intervals, each is of width $\frac{3-1}{8} = 0.25$ and ordinates occur at 1.00, 1.25, 1.50, 1.75, . . . 3.00

Corresponding values of $\frac{5}{x}$ are evaluated and shown in the following table.

x	1.00	1.25	1.50	1.75	2.00	2.25	2.50	2.75	3.00
$\frac{5}{x}$	5.0000	4.0000	3.3333	2.8571	2.5000	2.2222	2.0000	1.8182	1.6667

From equation (1), using the trapezoidal rule with 8 intervals:

$$\int_1^3 \frac{5}{x}\,dx \approx (0.25)\left[\frac{1}{2}(5.0000 + 1.6667) + 4.0000 + 3.3333 + 2.8571 + 2.5000 + 2.2222 + 2.0000 + 1.8182\right]$$
$$= \mathbf{5.516}, \text{ correct to 3 decimal places.}$$

(c) (i) From equation (5), using Simpson's rule with 4 intervals and the table of values in part (b) (i) above:

$$\int_1^3 \frac{5}{x}\,dx \approx \frac{1}{3}(0.5)\left[(5.0000+1.6667)+4(3.3333+2.0000)+2(2.5000)\right]$$

$$= \mathbf{5.500}, \text{ correct to 3 decimal places.}$$

(ii) From equation (5), using Simpson's rule with 8 intervals and the table of values in part (b) (ii) above:

$$\int_1^3 \frac{5}{x}\,dx \approx \frac{1}{3}(0.25)\left[(5.0000+1.6667+4(4.0000+2.8571+2.2222+1.8182)+2(3.3333+2.5000+2.0000)\right]$$

$$\approx \frac{1}{3}(0.25)[6.6667+43.5900+15.6666]$$

$$= \mathbf{5.494}, \text{ correct to 3 decimal places.}$$

(d) Percentage error = $\left(\dfrac{\text{Approximate value} - \text{true value}}{\text{true value}}\right) \times 100\%$, where true value = 5.493 from part (a).

With the trapezoidal rule using 4 intervals,

$$\text{percentage error} = \left(\frac{5.583 - 5.493}{5.493}\right) \times 100\% = \mathbf{1.638\%}$$

and using 8 intervals, percentage error = $\left(\dfrac{5.516 - 5.493}{5.493}\right) \times 100\%$ $= \mathbf{0.419\%}$

With Simpson's rule using 4 intervals,

$$\text{percentage error} = \left(\frac{5.500 - 5.493}{5.493}\right) \times 100\% = \mathbf{0.127\%}$$

and using 8 intervals, percentage error = $\left(\dfrac{5.494 - 5.493}{5.493}\right) \times 100\%$ $= \mathbf{0.018\%}$

Thus when evaluating $\displaystyle\int_1^3 \frac{5}{x}\,dx$ the following conclusions may be drawn from above:

(i) the larger the number of intervals chosen the more accurate is the result, and

(ii) Simpson's rule is more accurate than the trapezoidal rule when the same number of intervals are chosen.

Further problems on numerical integration may be found in the following section (5), problems 1 to 28.

---

5 Further problems

In problems 1 to 5, evaluate the definite integrals using the trapezoidal rule, giving the answers correct to 3 decimal places.

1. $\int_0^2 \frac{1}{1+\theta^2}\,d\theta$ (Use 8 intervals) [1.106]

2. $\int_1^4 3\ln 2x\,dx$ (Use 6 intervals) [13.827]

3. $\int_0^{\frac{\pi}{2}} \frac{1}{1+\sin x}\,dx$ (Use 6 intervals) [1.006]

4. $\int_0^{\frac{\pi}{2}} \sqrt{(\sin x)}\,dx$ (Use 5 intervals) [1.162]

5. $\int_0^{\pi} t\sin t\,dt$ (Use 8 intervals) [3.101]

In problems 6 to 10, evaluate the definite integrals using the mid-ordinate rule, giving the answers correct to 3 decimal places.

6. $\int_1^4 \sqrt{(x^2-1)}\,dx$ (Use 6 intervals) [6.735]

7. $\int_0^{\frac{\pi}{4}} \sqrt{(\cos^3\theta)}\,d\theta$ (Use 9 intervals) [0.674]

8. $\int_0^{1.5} e^{-\frac{1}{3}x^2}\,dx$ (Use 6 intervals) [1.197]

9. $\int_{0.4}^{2} \frac{dx}{1+x^4}$ (Use 8 intervals) [0.672]

10. $\int_0^{\frac{\pi}{2}} \frac{1}{1+\cos x}\,dx$ (Use 6 intervals) [0.997]

In problems 11 to 16, evaluate the definite integrals using Simpson's rule, giving the answers correct to 3 decimal places.

11. $\int_{\frac{\pi}{6}}^{\frac{\pi}{3}} \tan\theta \, d\theta$ (Use 6 intervals) [0.549]

12. $\int_{0}^{2} \frac{1}{1+x^3} \, dx$ (Use 8 intervals) [1.090]

13. $\int_{0}^{\frac{\pi}{3}} \sqrt{(1 - \frac{1}{3}\sin^2 x)} \, dx$ (Use 6 intervals) [0.994]

14. $\int_{1}^{3} \frac{\ln x}{x} \, dx$ (Use 10 intervals) [0.603]

15. $\int_{0}^{0.4} \frac{\sin\theta}{\theta} \, d\theta$ (Use 8 intervals) [0.380]

16. $\int_{0}^{\frac{\pi}{4}} \sqrt{(\sec x)} \, dx$ (Use 6 intervals) [0.831]

In problems 17 and 18 evaluate the definite integrals using (a) integration, (b) the trapezoidal rule, (c) the mid-ordinate rule, and (d) Simpson's rule. In each of the approximate methods give the answers correct to 3 decimal places.

17. $\int_{1}^{3} \frac{9}{x^2} \, dx$ (Use 8 intervals). [(a) 6 (b) 6.089 (c) 5.956 (d) 6.004]

18. $\int_{0}^{5} \sqrt{(3x+1)} \, dx$ (Use 10 intervals) [(a) 14 (b) 13.977 (c) 14.011 (d) 13.998]

In problems 19 to 21, evaluate the definite integrals using (a) the trapezoidal rule, (b) the mid-ordinate rule and (c) Simpson's rule. Use 6 intervals in each case and give answers correct to 3 decimal places.

19. $\int_{0}^{0.9} \sqrt{(1 - x^2)} \, dx$ [(a) 0.752 (b) 0.758 (c) 0.756]

20. $\int_{0.6}^{2.4} \sqrt{(1 - x^3)} \, dx$ [(a) 4.006 (b) 3.986 (c) 3.992]

21. $\int_{0}^{\frac{\pi}{2}} \frac{1}{\sqrt{(1 - \frac{1}{2}\sin^2\theta)}} \, d\theta$ [(a) 1.854 (b) 1.854 (c) 1.854]

22. A curve is given by the following values:

x	0	1.0	2.0	3.0	4.0	5.0	6.0
y	3	6	12	20	30	42	56

The area under the curve between $x = 0$ and $x = 6.0$ is given by $\int_0^{6.0} y \, dx$. Determine the approximate value of this definite integral, correct to 4 significant figures, using Simpson's rule. [138.3]

23. A function of x, $f(x)$, has the following values for corresponding values of x.

x	0	0.1	0.2	0.3	0.4	0.5	0.6
$f(x)$	0	0.0995	0.1960	0.2866	0.3684	0.4388	0.4952

Evaluate $\int_0^{0.6} f(x) \, dx$ using (a) the trapezoidal rule, and (b) Simpson's rule, giving answers correct to 3 decimal places. [(a) 0.164 (b) 0.164]

24. Use Simpson's rule to estimate $\int_1^3 y \, dx$ for the following pairs of (x, y) values.

x	1.00	1.25	1.50	1.75	2.00	2.25	2.50	2.75	3.00
y	0	0.2789	0.6082	0.9793	1.3863	1.8246	2.2907	2.7819	3.2958

[2.944]

25. A vehicle starts from rest and its velocity is measured every second for 6.0 seconds, with values as follows:

time t (s)	0	1.0	2.0	3.0	4.0	5.0	6.0
velocity v (m s^{-1})	0	1.2	2.4	3.7	5.2	6.0	9.2

The distance travelled in 6.0 seconds is given by $\int_0^{6.0} v \, dt$. Estimate this distance using Simpson's rule giving the answer correct to 3 significant figures. [22.7 m]

26. An alternating current i has the following values at equal intervals of 2.0×10^{-3} seconds.

time $\times 10^{-3}$ (s)	0	2.0	4.0	6.0	8.0	10.0	12.0
current i(A)	0	1.7	3.5	5.0	3.7	2.0	0

Charge q, in coulombs, is given by $q = \int_0^{12.0 \times 10^{-3}} i \, dt$. Use Simpson's rule to determine the approximate charge in the 12.0×10^{-3} second period. [32.8×10^{-3} C]

27. The velocity v of a body moving in a straight line at time t is given in the table below.

t(s)	0	0.5	1.0	1.5	2.0	2.5	3.0	3.5	4.0
v(m s^{-1})	0	0.07	0.13	0.22	0.27	0.32	0.34	0.31	0

The total distance travelled is given by $\int_0^{4.0} v \, dt$. Estimate the total distance travelled in 4.0 s using (a) the trapezoidal rule, and (b) Simpson's rule, giving the answers in centimetres. [(a) 83 cm (b) 86 cm]

28. Determine the value of $\int_0^2 \frac{x}{\sqrt{(2x^2 + 1)}} \, dx$ using integral calculus. Find also the percentage error introduced by estimating the definite integral by (a) the trapezoidal rule, (b) the mid-ordinate rule, and (c) Simpson's rule, using 4 intervals in each case. Give answers correct to 3 decimal places. [1; (a) −2.073%, (b) 1.063%, (c) 0.213%]

Chapter 7

Differential equations

1 Families of curves

A graph depicting the equations $y = 3x + 1$, $y = 3x + 2$, and $y = 3x - 4$ is shown in Fig. 1, and three parallel straight lines are seen to be the result. Equations of the form $y = 3x + c$, where c can have any numerical value, will produce an infinite number of parallel straight lines called a **family of curves.** A few of these can be seen in Fig. 1. Since $y = 3x + c$, $\frac{dy}{dx} = 3$, that is, the slope of every member of the family is 3. When additional information is given, for example, both the **general equation** $y = 3x + c$, and the member of the family passing through the point (2, 2), as shown as P in Fig. 1, then one particular member of the family is identified. The only line meeting both these conditions is the line $y = 3x - 4$. This is established by substituting $x = 2$ and $y = 2$ in the general equation $y = 3x + c$ and determining the value of c. Then, $2 = 3(2) + c$, giving $c = -6 + 2$, i.e. -4. Thus the **particular solution** is $y = 3x - 4$. Similarly, at point Q having coordinates $(-2, -4)$, the particular member of the family meeting both the conditions that it belongs to the family $y = 3x + c$ and that it passes through Q is the line $y = 3x + 2$. The additional information given, to enable a particular member of a family to be selected, is called the **boundary conditions.**

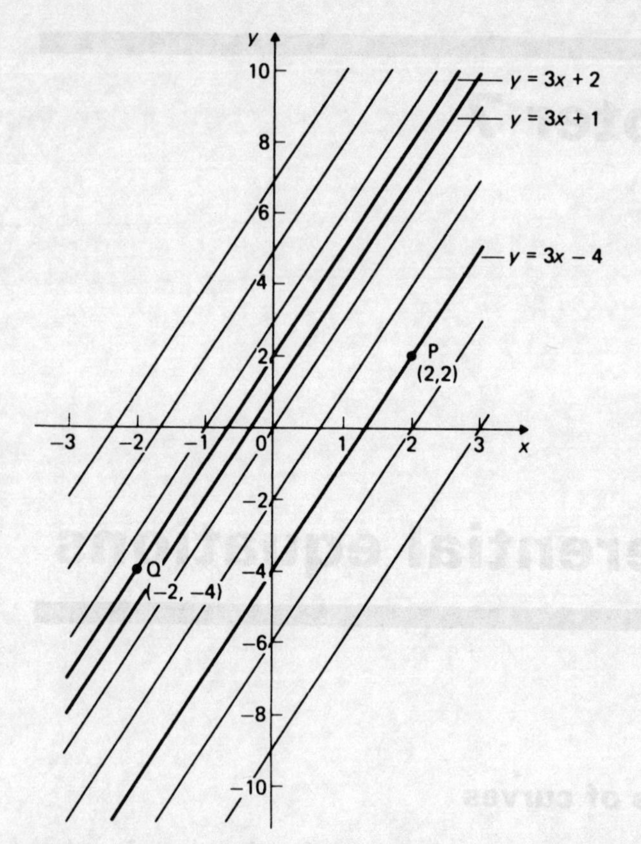

Fig. 1 Some members of the family of curves satisfying the equation $y = 3x + c$

The equation, $\frac{dy}{dx} = 3$, is called a **differential equation** since it contains a differential coefficient. It is also called a **first-order** differential equation, since it contains the first differential coefficient only, and has no differential coefficients such as $\frac{d^2y}{dx^2}$ or higher orders.

Another family of an infinite number of curves is produced by drawing a graph depicting the equations $y = 2x^2 + c$. Two of the curves in the family are $y = 2x^2$ (when $c = 0$) and $y = 2x^2 - 12$ (when $c = -12$) and these curves, together with others belonging to the family, are shown in Fig. 2.

The slope at any point of these curves is found by differentiating $y = 2x^2 + c$ and is given by $\frac{dy}{dx} = 4x$, i.e. the gradient of all of the curves is given by 4 times the value of the abscissa at every point. When boundary con-

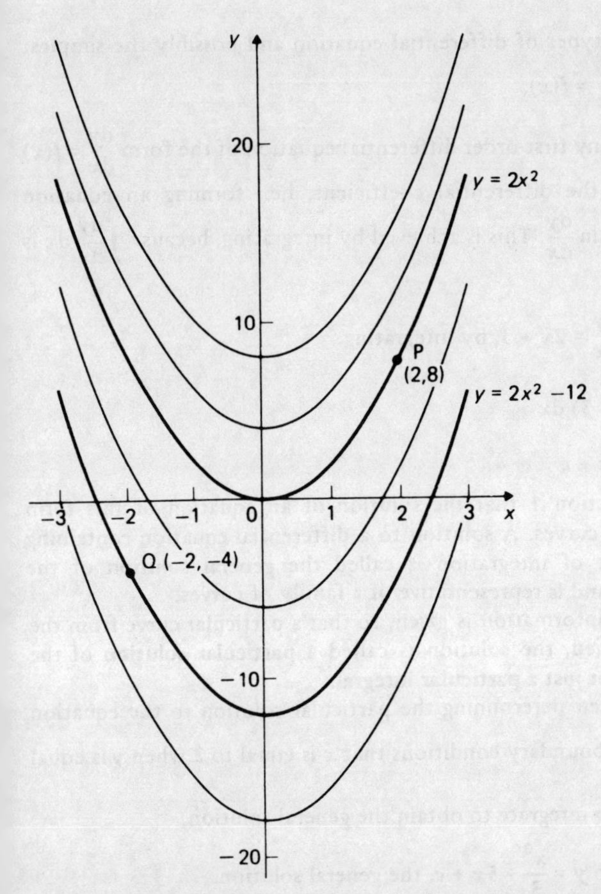

Fig. 2 Some members of the family of curves satisfying the equation $y = 2x^2 + c$

ditions are stated, particular curves can be identified. For example, the curve belonging to the family of curves $y = 2x^2 + c$ and which passes through point P, having coordinates (2, 8), is obtained by substituting $x = 2$ and $y = 8$ in the general equation. This gives $8 = 2(2)^2 + c$, i.e. $c = 0$, and hence the curve is $y = 2x^2$. Similarly, the curve satisfying the general equation and passing through the point Q = (−2, −4) is $y = 2x^2 - 12$.

2 The solution of differential equations of the form $\frac{dy}{dx} = f(x)$

Differential equations are used extensively in science and engineering.

There are many types of differential equation and possibly the simplest type is of the form $\frac{dy}{dx} = f(x)$.

The solution of any first-order differential equation of the form $\frac{dy}{dx} = f(x)$ involves eliminating the differential coefficient, i.e., forming an equation which does not contain $\frac{dy}{dx}$. This is achieved by integrating, because $\int \frac{dy}{dx}\,dx$ is equal to y.

For example, when $\frac{dy}{dx} = 2x + 3$, by integrating:

$$\int \frac{dy}{dx}\,dx = \int (2x + 3)\,dx$$

giving $\quad y = x^2 + 3x + c$

It was shown in Section 1 that the solution of an equation of this form produces a family of curves. A solution to a differential equation containing an arbitrary constant of integration is called the **general solution** of the differential equation and is representative of a family of curves.

When additional information is given, so that a particular curve from the family can be identified, the solution is called a **particular solution** of the differential equation or just a **particular integral.**

For example, when determining the particular solution to the equation $\frac{dy}{dx} = x^2 + 5$, given the boundary conditions that x is equal to 2 when y is equal to 5, the first step is to integrate to obtain the general solution.

When $\frac{dy}{dx} = x^2 + 5$, then $y = \frac{x^3}{3} + 5x + c$, the general solution.

Using the information given, i.e. substituting for x and y in the general solution, gives:

$$5 = \frac{(2)^3}{3} + 5(2) + c$$

and so $c = 5 - \frac{8}{3} - 10 = -7\frac{2}{3}$

Hence the particular solution is

$$y = \frac{x^3}{3} + 5x - 7\tfrac{2}{3}.$$

Worked problems on the solution of differential equations of the form $\frac{dy}{dx} = f(x)$

Problem 1. Find the general solutions to the equations:

(a) $\frac{dy}{dx} + \frac{5}{x} = 4x$

(b) $x\frac{dy}{dx} = 3x^3 - 4x^2 + 5x$

(c) $5\frac{dM}{d\theta} = 3e^{\theta} - 4e^{-\theta}$

(d) $\frac{ds}{dt} = u + at$, where u and a are constants

(e) $\frac{di}{dt} = \omega I_m \cos \omega t$, where ω and I_m are constants

Each of these equations is of the form $\frac{dy}{dx} = f(x)$ and the general solution can be obtained by integration.

(a) $\frac{dy}{dx} = 4x - \frac{5}{x}$

Integrating: $\int \frac{dy}{dx}\,dx = \int \left(4x - \frac{5}{x}\right) dx$

$$y = \frac{4x^2}{2} - 5 \ln x + c$$

i.e. $\quad y = 2x^2 - 5 \ln x + c$

(b) $x\frac{dy}{dx} = 3x^3 - 4x^2 + 5x$

Dividing throughout by x gives:

$$\frac{dy}{dx} = 3x^2 - 4x + 5$$

Integrating gives: $y = \frac{3x^3}{3} - \frac{4x^2}{2} + 5x + c$

i.e. $\quad y = x^3 - 2x^2 + 5x + c$

(c) $5\frac{dM}{d\theta} = 3e^{\theta} - 4e^{-\theta}$

$$\frac{dM}{d\theta} = \tfrac{1}{5}(3e^{\theta} - 4e^{-\theta})$$

Integrating gives: $M = \frac{1}{5}(3e^{\theta} + 4e^{-\theta}) + c$

(d) $\frac{ds}{dt} = u + at$

Integrating gives: $s = ut + \frac{1}{2}at^2 + c$

(e) $\frac{di}{dt} = \omega I_m \cos \omega t$

Integrating gives: $i = \frac{\omega I_m}{\omega} \sin \omega t + c$

i.e. $i = I_m \sin \omega t + c$

Problem 2. Find the particular solutions of the following equations satisfying the given boundary conditions:

(a) $\frac{dy}{dx} + x = 2$ and $y = 3$ when $x = 1$

(b) $x\frac{dy}{dx} = 3 - x^3$ and $y = 3\frac{2}{3}$ when $x = 1$

(c) $3\frac{dr}{d\theta} + \cos\theta = 0$ and $r = 5$ when $\theta = \frac{\pi}{2}$

(d) $\frac{dy}{dx} = e^x - 2\sin 2x$ and $y = 2$ when $x = \frac{\pi}{4}$

(a) $\frac{dy}{dx} = 2 - x$

Integrating gives: $y = 2x - \frac{x^2}{2} + c$, the general solution.

Substituting the boundary conditions $y = 3$ and $x = 1$ to evaluate c gives:

$$3 = 2(1) - \frac{(1)^2}{2} + c$$

i.e. $c = 1\frac{1}{2}$

Hence the particular solution is $\boldsymbol{y = 2x - \frac{x^2}{2} + 1\frac{1}{2}}$

(b) $x\frac{dy}{dx} = 3 - x^3$

Dividing throughout by x to express the equation in the form $\frac{dy}{dx} = f(x)$ gives:

$$\frac{dy}{dx} = \frac{3}{x} - x^2$$

Integrating gives: $y = 3\ln x - \frac{x^3}{3} + c$, the general solution.

Substituting the boundary conditions gives: $3\frac{2}{3} = 3\ln 1 - \frac{(1)^3}{3} + c$

i.e. $c = 4$

Hence the particular solution is $\boldsymbol{y = 3\ln x - \frac{x^3}{3} + 4}$

(c) $3\frac{dr}{d\theta} + \cos\theta = 0$

$$\frac{dr}{d\theta} = -\frac{1}{3}\cos\theta$$

Integrating gives: $r = -\frac{1}{3}\sin\theta + c$, the general solution.

Substituting the boundary conditions gives: $5 = -\frac{1}{3}\sin\frac{\pi}{2} + c$

i.e. $c = 5\frac{1}{3}$

Hence the particular solution is: $r = -\frac{1}{3}\sin\theta + 5\frac{1}{3}$

(d) $\frac{dy}{dx} = e^x - 2\sin 2x$

Integrating gives: $y = e^x + \cos 2x + c$, the general solution.

Substituting the boundary conditions gives: $2 = e^{\pi/4} + \cos\frac{\pi}{2} + c$

i.e. $c = 2 - e^{\pi/4}$

Hence the particular solution is $y = e^x + \cos 2x + 2 - e^{\pi/4}$

Expressed in this form, the true value of y is stated. The value of $e^{\pi/4}$ is 2.193 3 correct to 4 decimal places and the result can be expressed as $y = e^x + \cos 2x - 0.193\ 3$, correct to 4 decimal places. However, when a result can be accurately expressed in terms of e or π, then it is usually better to leave the result in this form, unless the problem specifies the accuracy required.

Further problems on the solution of differential equations of the form $\frac{dy}{dx} = f(x)$ *may be found in Section 4, Problems 1–30, page 146.*

3 The solution of differential equations of the form $\frac{dQ}{dt} = kQ$

The natural laws of growth and decay are of the form $y = Ae^{kx}$, where A and k are constants. For such a law to apply, the rate of change of a variable must be proportional to the variable itself. This can be shown by differentiation. Since

$$y = Ae^{kx}$$

$$\frac{dy}{dx} = Ake^{kx}, \text{ i.e. } \frac{dy}{dx} = kAe^{kx}$$

But Ae^{kx} is equal to y. Hence, $\frac{dy}{dx} = ky$

Three of the natural laws are shown below and all such laws can be shown to be of a similar form.

(i) For linear expansion, the amount by which a rod expands when heated depends on the length of the rod, that is, the increase of length with respect to temperature is proportional to the length of the rod. Thus, mathematically:

$$\frac{dl}{d\theta} = kl \text{ and the law is } l = l_0 e^{k\theta}$$

(ii) For Newton's law of cooling, the fall of temperature with respect to time is proportional to the excess of its temperature above that of its surroundings, i.e.

$$\frac{d\theta}{dt} = -k\theta \text{ and the law is } \theta = \theta_0 e^{-kt}$$

(iii) In electrical work, when current decays in a circuit containing resistance and inductance connected in series, the change of current with respect to time is proportional to the current flowing at any instant,

i.e. $\dfrac{di}{dt} = ki$ and the law is $i = Ae^{kt}$ where $k = -\dfrac{1}{T}$ and t is the time constant of the circuit.

These are just some of many examples of natural or exponential laws. In general, differential equations of the form $\dfrac{dQ}{dt} = kQ$ depict natural laws and the solutions are always of the form $Q = Ae^{kt}$. This can be shown as follows:

Since $\dfrac{dQ}{dt} = kQ, \dfrac{dQ}{Q} = k\, dt$

Integrating: $\displaystyle\int \frac{dQ}{Q} = \int k\, dt$

i.e. $\ln Q = kt + c$

By the definition of a logarithm, if $y = e^x$ then $x = \ln y$, i.e. if $\ln y = x$, then $y = e^x$. It follows that when $\ln Q = kt + c$

$$Q = e^{(kt+c)}$$

By the laws of indices, $e^a e^b = e^{(a+b)}$ and applying this principle, gives:

$$Q = e^{kt} e^c$$

But e^c is a constant, say A, thus

$$Q = Ae^{kt}$$

Checking by differentiation:

when $Q = Ae^{kt}$

$$\frac{dQ}{dt} = kAe^{kt} = kQ$$

Hence Ae^{kt} is a solution to the differential equation $\dfrac{dQ}{dt} = kQ$.

Thus the general solution of any differential equation of the form

$$\frac{dQ}{dt} = kQ \text{ is } Q = Ae^{kt}$$

and when boundary conditions are given, the particular solution can be obtained, as shown in the worked problems following.

Worked problems on the solution of equations of the form $\frac{dQ}{dt} = kQ$

Problem 1. Solve the equation $\frac{dy}{dx} = 6y$ given that $y = 3$ when $x = 0.5$.

Since $\frac{dy}{dx} = 6y$ is of the form $\frac{dQ}{dt} = kQ$, the solution to the general equation will be of the form $Q = Ae^{kt}$, i.e. $y = Ae^{6x}$

Substituting the boundary conditions gives: $3 = Ae^{6(0.5)}$

i.e. $$A = \frac{3}{e^3} = 0.1494$$

Hence the particular solution is $y = \frac{3}{e^3}e^{6x} = 3e^{3(2x-1)}$ or $0.1494e^{6x}$

Problem 2. Determine the particular solutions of the following equations and their given boundary conditions, expressing the values of the constants correct to 4 significant figures:

(a) $\frac{dM}{di} - 4M = 0$ and $M = 5$ when $i = 1$

(b) $\frac{1}{15}\frac{dl}{dm} + \frac{l}{4} = 0$ and $l = 15.41$ when $m = 0.714\,3$

(a) Rearranging the equation into the form $\frac{dQ}{dt} = kQ$ gives:

$$\frac{dM}{di} = 4M$$

The general solution is of the form $Q = Ae^{kt}$, giving

$M = Ae^{4i}$

Substituting the boundary conditions gives: $5 = Ae^{(4)(1)}$

i.e. $$A = \frac{5}{e^4} = 0.091\,58$$

Hence the particular solution is $M = 0.091\,58\,e^{4i}$

(b) Writing the equation in the form $\frac{dQ}{dt} = kQ$ gives:

$$\frac{dl}{dm} = -\frac{15}{4}l$$

The general solution is $I = Ae^{-\frac{15}{4}m}$

Substituting the boundary conditions gives: $15.41 = Ae^{(-\frac{15}{4})(0.714\,3)}$

i.e. $A = 224.4$

Hence the particular solution is $\mathbf{I = 224.4\,e^{-3.750m}}$

Problem 3. The decay of current in an electrical circuit containing resistance R ohms and inductance L henrys in series is given by $L\frac{di}{dt} + Ri = 0$, where i is the current flowing at time t seconds. Determine the general solution of the equation. In such a circuit, R is 5 kΩ, L is 3 henrys and the current falls to 5 A in 0.7 ms. Determine how long it will take for the current to fall to 2 amperes. Express your answer correct to 2 significant figures.

Since $\frac{di}{dt} = -\frac{R}{L}i$, then the general solution of the equation is

$i = Ae^{-\frac{Rt}{L}}$

By substituting the given values of R, L, i and t, the value of constant A is determined, i.e.

$$5 = Ae^{\left(\frac{-5\times10^3\times0.7\times10^{-3}}{3}\right)}$$

$$= Ae^{-\frac{3.5}{3}}$$

giving $A = 16.056$

To determine the time for i to fall to 2 A, substituting in the general solution for i, A, R and L gives

$$2 = 16.056\,e^{\left(\frac{-5\times10^3\times t}{3}\right)}$$

$$= 16.056\,e^{\left(\frac{-5t}{3}\right)}$$ when t is stated in milliseconds.

Thus, $e^{\left(-\frac{5t}{3}\right)} = \frac{2}{16.056} = 0.124\,56$,

and taking natural logarithms, gives:

$-\frac{5t}{3}\ln e = \ln 0.124\,56$

But $\ln e = 1$, hence $t = -\frac{3}{5}\ln 0.124\,56 = 1.25$ ms

i.e. **the time for i to fall to 2 A is 1.3 ms**, correct to 2 significant figures.

Problem 4. A copper conductor heats up to 50°C when carrying a current of 200 A. If the temperature coefficient of linear expansion, α_0, for copper is 17×10^{-6}/°C at 0°C and the equation relating temperature θ with length l is $\frac{dl}{d\theta} = \alpha l$, find the increase in length of the conductor at 50°C correct to the nearest centimetre, when l is 1 000 m at 0°C.

Since $\frac{dl}{d\theta} = \alpha l$, then $l = Ae^{\alpha\theta}$, the general solution.

But l is 1 000 when θ is 0°C, and substituting these values in the general solution of the equation gives:

$1\,000 = Ae^0$, i.e. $A = 1\,000$

Substituting for A, α and θ in the general equation, gives

$l = 1\,000\ e^{(17 \times 10^{-6} \times 50)} = 1\,000.850$ m

i.e. **the increase in length of the conductor at 50°C is 85 cm**, correct to the nearest centimetre.

Problem 5. The rate of cooling of a body is proportional to the excess of its temperature above that of its surrounding, θ°C.

The equation is: $\frac{d\theta}{dt} = k\theta$, where k is a constant.

A body cools from 90°C to 70°C in 3.0 minutes at a surrounding temperature of 15°C. Determine how long it will take for the body to cool to 50°C.

The general solution of the equation $\frac{d\theta}{dt} = k\theta$ is $\theta = Ae^{kt}$.

Letting the temperature 90°C correspond to a time t of zero gives an excess of body temperature above the surroundings of (90 – 15).

Hence, $(90 - 15) = Ae^{(k)(0)}$, i.e. $A = 75$.

3.0 minutes later, the general solution becomes:

$(70 - 15) = 75\ e^{(k)(3)}$

i.e. $e^{3k} = \frac{55}{75}$

Taking natural logarithms,

$3k = \ln\frac{55}{75}$, $k = \frac{1}{3}\ln\frac{55}{75}$

i.e. $k = -0.103\,38$

At 50°C, $(50 - 15) = 75\ e^{-0.103\,38t}$

$\frac{35}{75} = e^{-0.103\,38t}$

Taking natural logarithms gives:

$t = -\frac{1}{0.103\,38}\ln\frac{35}{75}$

$= 7.37$

That is, **the time for the body to cool to 50°C is 7.37 minutes**, or 7 minutes 22 seconds, correct to the nearest second.

Problem 6. The rate of decay of a radioactive material is given by $\frac{dN}{dt} = -\lambda N$ where λ is the decay constant and N the number of radioactive atoms disintegrating per second. Determine the half-life of a zinc isotope, taking the decay constant as 2.22×10^{-4} atoms per second.

The half-life of an element is the time for N to become one-half of its original value. Since $\frac{dN}{dt} = -\lambda N$, then applying the general solution to this equation gives:

$N = Ae^{-\lambda t}$, where the constant A represents the original number of radioactive atoms present since $N = A$ when $t = 0$. For half-life conditions, the ratio $\frac{N}{A}$ is $\frac{1}{2}$, hence

$$\tfrac{1}{2} = e^{-\lambda t} = e^{-2.22 \times 10^{-4} t}$$

Thus, $\ln \tfrac{1}{2} = -2.22 \times 10^{-4} t$

i.e. $t = -\frac{1}{2.22 \times 10^{-4}} \ln 0.5$

$= 3\,122$ seconds or 52 minutes, 2 seconds.

Thus, the half-life is 52 minutes, correct to the nearest minute.

Further problems on the solution of equations of the form $\frac{dQ}{dt} = kQ$ *may be found in the following section (4), Problems 31–41, page 148.*

4 Further problems

Solution of equations of the form $\frac{dy}{dx} = f(x)$

In Problems 1–15, find the general solutions of the equations.

1. $\frac{dy}{dx} = 3x - \frac{4}{x^2}$ $\quad\left[y = \frac{3x^2}{2} + \frac{4}{x} + c\right]$
2. $\frac{dy}{dx} + 3 = 4x^2$ $\quad\left[y = \frac{4x^3}{3} - 3x + c\right]$
3. $3\frac{dy}{dx} + \frac{2}{\sqrt{x}} = 5\sqrt{x}$ $\quad[y = \tfrac{2}{3}\sqrt{x}(\tfrac{5}{3}x - 2) + c]$
4. $\frac{du}{dV} - \frac{1}{V} = 4$ $\quad[u = 4V + \ln V + c]$
5. $6 - 5\frac{dy}{dx} = \frac{1}{x-2}$ $\quad[y = \tfrac{1}{5}(6x - \ln(x-2)) + c]$
6. $2x^2 - \frac{3}{x} + 4\frac{dy}{dx} = 0$ $\quad\left[y = \tfrac{1}{4}\left(3\ln x - 2\frac{x^3}{3}\right) + c\right]$

7. $\dfrac{di}{d\theta} = \cos\theta$ $\quad [i = \sin\theta + c]$

8. $6\dfrac{dV}{dt} = 4\sin\left(100t + \dfrac{\pi}{6}\right)$ $\quad \left[V = -\dfrac{1}{150}\cos\left(100t + \dfrac{\pi}{6}\right) + c\right]$

9. $\dfrac{di}{dt} - \dfrac{t}{10} + 140 = 0$ $\quad \left[i = \dfrac{t^2}{20} - 140t + c\right]$

10. $3\dfrac{dv}{dt} + 0.7t^2 - 1.4 = 0$ $\quad [v = \frac{1}{9}(4.2t - 0.7t^3) + c]$

11. $\dfrac{dy}{d\theta} = 3e^{\theta} - \dfrac{4}{e^{2\theta}}$ $\quad \left[y = 3e^{\theta} + \dfrac{2}{e^{2\theta}} + c\right]$

12. $\dfrac{dV}{dx} = 3x - \dfrac{5}{x} - \sec^2 x$ $\quad \left[V = \dfrac{3x^2}{2} - 5\ln x - \tan x + c\right]$

13. $\frac{1}{2}\dfrac{dy}{dx} + 2x^{\frac{1}{2}} = e^{\frac{x}{2}}$ $\quad \left[y = 4\left(e^{\frac{x}{2}} - \dfrac{2x^{\frac{3}{2}}}{3}\right) + c\right]$

14. $x\dfrac{dy}{dx} = 2 - 3x^2$ $\quad \left[y = 2\ln x - \dfrac{3x^2}{2} + c\right]$

15. $\dfrac{dM}{d\theta} = \frac{1}{2}\sin 3\theta - \frac{1}{3}\cos 2\theta$ $\quad [M = -\frac{1}{6}(\cos 3\theta + \sin 2\theta) + c]$

In Problems 16–25, determine the particular solutions of the differential equations for the boundary conditions given.

16. $x\dfrac{dy}{dx} - 2 = x^3$ and $y = 1$ when $x = 1$. $\quad \left[y = 2\ln x + \dfrac{x^3}{3} + \frac{2}{3}\right]$

17. $x\left(x - \dfrac{dy}{dx}\right) = 3$ and $y = 2$ when $x = 1$. $\quad \left[y = \dfrac{x^2}{2} - 3\ln x + 1\frac{1}{2}\right]$

18. $\dfrac{ds}{dt} - 4t^2 = 9$ and $s = 27$ when $t = 3$. $\quad \left[s = 9t + \dfrac{4t^3}{3} - 36\right]$

19. $e^{-p}\dfrac{dq}{dp} = 5$ and $q = 2.718$ when $p = 0$. $\quad [q = 5e^{p} - 2.282]$

20. $3 - \dfrac{dy}{dx} = e^{2x} - 2e^{x}$ and $y = 7$ when $x = 0$.
$[y = 3x - \frac{1}{2}e^{2x} + 2e^{x} + 5\frac{1}{2}]$

21. $\dfrac{dy}{d\theta} - \sin 3\theta = 5$ and $y = \dfrac{5\pi}{6}$ when $\theta = \dfrac{\pi}{6}$. $\quad [y = 5\theta - \frac{1}{3}\cos 3\theta]$

22. $3\sin\left(2\theta - \dfrac{\pi}{3}\right) + 4\dfrac{dv}{d\theta} = 0$ and $v = 3.7$ when $\theta = \dfrac{2\pi}{3}$.
$\left[v = \frac{3}{8}\left(\cos\left(2\theta - \dfrac{\pi}{3}\right)\right) + 4.075\right]$

23. $\frac{1}{6}\dfrac{dM}{d\theta} + 1 = \sin\theta$ and $M = 3$ when $\theta = \pi$.
$[M = 3\{(2\pi - 1) - 2\cos\theta - 2\theta\}]$

24. $\dfrac{2}{(u+1)^2} = 4 - \dfrac{dz}{du}$ and $z = 14$ when $u = 5$. $\quad \left[z = 4u + \dfrac{2}{u+1} - 6\frac{1}{3}\right]$

25. $\dfrac{1}{2e^{x}} + 4 = x - 3\dfrac{dy}{dx}$ and $y = 3$ when $x = 0$.
$\left[y = \frac{1}{3}\left\{\dfrac{x^2}{2} + \dfrac{1}{2e^{x}} - 4x\right\} + 2\frac{5}{6}\right]$

26. The bending moment of a beam, M, and shear force F are related by the equation $\frac{dM}{dx} = F$, where x is the distance from one end of the beam. Determine M in terms of x when $F = -w(l - x)$ where w and l are constants, and $M = \frac{1}{2}wl^2$ when $x = 0$. $[M = \frac{1}{2}w(l - x)^2]$
27. The angular velocity ω of a flywheel of moment of inertia I is given by $I\frac{d\omega}{dt} + N = 0$, where N is a constant. Determine ω in terms of t given that $\omega = \omega_0$ when $t = 0$. $\left[\omega = \omega_0 - \frac{Nt}{I}\right]$
28. The gradient of a curve is given by $\frac{dy}{dx} = 2x - \frac{x^2}{3}$. Determine the equation of the curve if it passes through the point $x = 3$, $y = 4$.
$\left[y = x^2 - \frac{x^3}{9} - 2\right]$
29. The acceleration of a body a is equal to its rate of change of velocity, $\frac{dv}{dt}$. Determine an equation for v in terms of t given that the velocity is u when $t = 0$. $[v = u + at]$
30. The velocity of a body v, is equal to its rate of change of distance, $\frac{dx}{dt}$. Determine an equation for x in terms of t given $v = u + at$, where u and a are constants and $x = 0$ when $t = 0$. $[x = ut + \frac{1}{2}at^2]$

Solution of equations of the form $\frac{dQ}{dt} = kQ$

In Problems 31–33 determine the general solutions to the equations.

31. $\frac{dp}{dq} = 9p$ $[p = Ae^{9q}]$
32. $\frac{dm}{dn} + 5m = 0$ $[m = Ae^{-5n}]$
33. $\frac{1}{6}\frac{dw}{dx} + \frac{3}{5}w = 0$ $[w = Ae^{-\frac{18}{5}x}]$

In Problems 34–36 determine the particular solutions to the equations, expressing the values of the constants correct to 3 significant figures.

34. $\frac{dQ}{dt} = 15.0Q$ and $Q = 7.3$ when $t = 0.015$. $[Q = 5.83\,e^{15.0t}]$
35. $\frac{1}{7}\frac{dl}{dm} - \frac{1}{3}l = 0$ and $l = 1.7 \times 10^4$ when $m = 3.4 \times 10^{-2}$.
$[l = 1.57 \times 10^4\,e^{2.33m}]$
36. $0.741\frac{dy}{dx} + 0.071\,y = 0$ and $y = 73.4$ when $x = 15.7$
$[y = 330\,e^{-0.0958x}]$

37. The difference in tension, T newtons, between two sides of a belt when in contact with a pulley over an angle of θ radians and when it is on the point of slipping, is given by $\frac{dT}{d\theta} = \mu T$, where μ is the coefficient of friction between the material of the belt and that of the pulley at the point of slipping. When $\theta = 0$ radians, the tension is 170 N and the coefficient of friction as slipping starts is 0.31. Determine the tension at the point of slipping when θ is $\frac{5\pi}{6}$ radians. Also determine the angle of lap in degrees, to give a tension of 340 N just before slipping starts.
[383 N, 128°]

38. The charge Q coulombs at time t seconds for a capacitor of capacitance C farads when discharging through a resistance of R ohms is given by:
$$R\frac{dQ}{dt} + \frac{Q}{C} = 0$$
A circuit contains a resistance of 500 kilohms and a capacitance of 8.7 microfarads, and after 147 milliseconds the charge falls to 7.5 coulombs. Determine the initial charge and the charge after one second, correct to 3 significant figures. [7.76 C, 6.17 C]

39. The rate of decay of a radioactive substance is given by $\frac{dN}{dt} = -\lambda N$, where λ is the decay constant and N the number of radioactive atoms disintegrating per second. Determine the half-life of radium in years (i.e. the time for N to become one-half of its original value) taking the decay constant for radium as 1.36×10^{-11} atoms per second and assuming a '365-day' year. [1 616 years]

40. The variation of resistance, R ohms, of a copper conductor with temperature, θ°C, is given by $\frac{dR}{d\theta} = \alpha R$, where α is the temperature coefficient of resistance of copper. Taking α as 39×10^{-4} per °C, determine the resistance of a copper conductor at 30°C, correct to 4 significant figures, when its resistance at 80°C is 57.4 ohms. [47.23 ohms]

41. The rate of growth of bacteria is directly proportional to the amount of bacteria present. Form a differential equation for the rate of growth when n is the number of bacteria at time t seconds. If the number of bacteria present at $t = 0$ is n_0, solve the equation. When the number of bacteria doubles in one hour, determine by how many times it will have increased in twelve hours. [$n = n_0 e^{kt}$, 2^{12}]

Appendices

Appendix A The binomial expansion

The general binomial expansion of $(a + b)^n$, where n is any positive integer, is given by:

$$(a + b)^n = a^n + n\,a^{n-1}\,b + \frac{n\,(n-1)}{(1)\,(2)}\,a^{n-2}\,b^2 + \frac{n\,(n-1)(n-2)}{(1)\,(2)\,(3)}\,a^{n-3}\,b^3 + \ldots,$$

Appendix B Trigonometrical compound angles

1. Compound angles

Angles such as $(A + B)$ or $(A - B)$ are called **compound angles** since they are the sum or difference of two angles A and B. Each expression of a compound angle has two components. The trigonometrical ratios of compound angles may be expressed in terms of their two component angles. These are often called the **addition and subtraction formulae** and are true for all values of A and B. It may be shown that:

$$\sin(A + B) = \sin A \cos B + \cos A \sin B \qquad \ldots(1)$$

$$\sin(A - B) = \sin A \cos B - \cos A \sin B \qquad \ldots(2)$$

$$\cos(A + B) = \cos A \cos B - \sin A \sin B \qquad \ldots(3)$$

$$\cos(A - B) = \cos A \cos B + \sin A \sin B \qquad \ldots(4)$$

2. Changing sums or differences of sines and cosines into products of sines and cosines

From equations (1) and (2) in section 1,

$$\sin(A + B) + \sin(A - B) = \sin A \cos B + \cos A \sin B + \sin A \cos B - \cos A \sin B$$

$$= 2 \sin A \cos B.$$

In the compound angle formulae let $A + B = x$
and $A - B = y$.

Then $A = \dfrac{x + y}{2}$ and $B = \dfrac{x - y}{2}$.

Then, instead of $\sin(A + B) + \sin(A - B) = 2 \sin A \cos B$ we have

$$\sin x + \sin y = 2 \sin \left(\frac{x+y}{2}\right) \cos \left(\frac{x-y}{2}\right) \qquad \ldots (5)$$

Similarly,

$$\sin x - \sin y = 2 \cos \left(\frac{x+y}{2}\right) \sin \left(\frac{x-y}{2}\right) \qquad \ldots (6)$$

$$\cos x + \cos y = 2 \cos \left(\frac{x+y}{2}\right) \cos \left(\frac{x-y}{2}\right) \qquad \ldots (7)$$

and

$$\cos x - \cos y = -2 \sin \left(\frac{x+y}{2}\right) \sin \left(\frac{x-y}{2}\right) \qquad \ldots (8)$$

3. Double angles

If in equation (1) of section 1, $A = B$ then

$$\sin 2A = 2 \sin A \cos A \qquad \ldots (9)$$

Similarly, if in equation 3, of section 1, $A = B$ then

$$\cos 2A = \cos^2 A - \sin^2 A \qquad \ldots (10)$$

There are two further formulae for $\cos 2A$. Since $\sin^2 A + \cos^2 A = 1$ then $\sin^2 A = 1 - \cos^2 A$. Thus

$$\cos 2A = \cos^2 A - (1 - \cos^2 A) = 2 \cos^2 A - 1 \qquad \ldots (11)$$

Similarly, $\cos 2A = (1 - \sin^2 A) - \sin^2 A = 1 - 2 \sin^2 A \qquad \ldots (12)$

Note that since $\sin 2A = 2 \sin A \cos A$, then $\sin 4A = 2 \sin 2A \cos 2A$ and since $\cos 2A = 2 \cos^2 A - 1$, then $\cos 6A = 2 \cos^2 3A - 1$, and so on.

Appendix C Approximate methods for finding areas of irregular figures

1. Trapezoidal rule

For Fig. 1(a), area of ABCD $= d\left[\left(\frac{y_1 + y_7}{2}\right) + y_2 + y_3 + y_4 + y_5 + y_6\right]$

i.e. the trapezoidal rule states that the area of an irregular figure is given by:
Area = (width of interval) [$\frac{1}{2}$ (first + last ordinates) + sum of remaining ordinates]

2. Mid-ordinate rule

For Fig. 1(b), area of PQRS $= d\,(y_1 + y_2 + y_3 + y_4 + y_5 + y_6)$,
i.e. the mid-ordinate rule states that the area of an irregular figure is given by:
Area = (width of interval) (sum of mid-ordinate)

3. Simpson's rule

To find an area such as ABCD of Fig. 1(a) the base AD **must** be divided into an **even** number of strips of equal width d, thus producing an **odd** number of ordinates, in this case 7.

$$\text{Area of ABCD} = \frac{d}{3}\,[(y_1 + y_7) + 4\,(y_2 + y_4 + y_6) + 2\,(y_3 + y_5)]$$

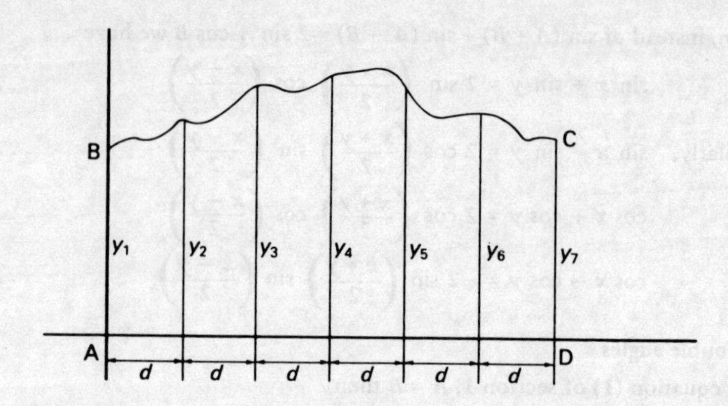

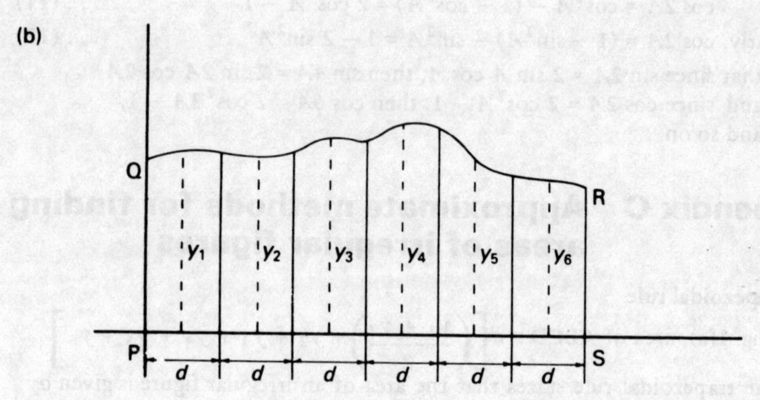

Fig. 1

i.e. Simpson's rule states that the area of an irregular figure is given by:

$$\text{Area} = \tfrac{1}{3}(\text{width of interval})\left[\begin{pmatrix}\text{first + last}\\ \text{ordinate}\end{pmatrix} + 4\begin{pmatrix}\text{sum of even}\\ \text{ordinates}\end{pmatrix} + 2\begin{pmatrix}\text{sum of remaining}\\ \text{odd ordinates}\end{pmatrix}\right]$$

When estimating areas of irregular figures, Simpson's rule is generally regarded as the most accurate of the approximate methods available.

Differential coefficients of common functions

1. Differential coefficient of ax^n

Let $f(x) = ax^n$
then $f(x + \delta x) = a(x + \delta x)^n$

By definition, $f'(x) = \lim_{\delta x \to 0} \left\{ \frac{f(x + \delta x) - f(x)}{\delta x} \right\}$

$= \lim_{\delta x \to 0} \left\{ \frac{a(x + \delta x)^n - ax^n}{\delta x} \right\}$

$a(x + \delta x)^n$ may be expanded using the binomial expansion (see Appendix A).

$$a(x + \delta x)^n = a \left[x^n + nx^{n-1}\,\delta x + \frac{n(n-1)}{(1)(2)} x^{n-2}\,(\delta x)^2 + \ldots \right]$$

$$= ax^n + an\,x^{n-1}\,\delta x + \frac{a\,n(n-1)}{(1)(2)} x^{n-2}\,(\delta x)^2 + \ldots$$

$$a(x + \delta x)^n - ax^n = an\,x^{n-1}\,\delta x + \frac{a\,n(n-1)}{(1)(2)} x^{n-2}\,(\delta x)^2 + \ldots$$

$$\frac{a(x + \delta x)^n - ax^n}{\delta x} = an\,x^{n-1} + \frac{a\,n(n-1)}{(1)(2)} x^{n-2}\,\delta x + \ldots$$

$$f'(x) = \lim_{\delta x \to 0} \left\{ an\,x^{n-1} + \frac{a\,n(n-1)}{(1)(2)} x^{n-2}\,\delta x + \ldots \right\}$$

i.e. $f'(x) = an\ x^{n-1}$, since all subsequent terms in the bracket will contain δx raised to some power and will become zero when a limiting value is taken. This result is true for all values of n, whether they are positive, negative or fractional.

Hence when $y = f(x) = ax^n$, $f'(x) = an\ x^{n-1}$

2. Differential coefficient of sin x

Let $f(x) = \sin x$
then $f(x + \delta x) = \sin (x + \delta x)$

By definition, $f'(x) = \lim_{\delta x \to 0} \left\{ \frac{f(x + \delta x) - f(x)}{\delta x} \right\}$

$= \lim_{\delta x \to 0} \left\{ \frac{\sin (x + \delta x) - \sin x}{\delta x} \right\}$

Now $\sin x - \sin y = 2 \cos \left[\frac{x + y}{2} \right] \sin \left[\frac{x - y}{2} \right]$ (see Appendix B).

Hence $\sin (x + \delta x) - \sin x = 2 \cos \left[\frac{(x + \delta x) + x}{2} \right] \sin \left[\frac{(x + \delta x) - x}{2} \right]$

$$= 2 \cos\left[x + \frac{\delta x}{2}\right] \sin\left[\frac{\delta x}{2}\right]$$

$$\frac{\sin(x + \delta x) - \sin x}{\delta x} = \frac{2 \cos\left[x + \frac{\delta x}{2}\right] \sin\left[\frac{\delta x}{2}\right]}{\delta x}$$

$$= \cos\left[x + \frac{\delta x}{2}\right] \frac{\sin\left[\frac{\delta x}{2}\right]}{\left[\frac{\delta x}{2}\right]}$$

When δx is small, $\sin \delta x \simeq \delta x$
(For example, if $\delta x = 1°$, $\sin 1° = 0.017\,5$ and $1°$ in radians $= 0.017\,5$.)

Thus when δx is small, say, less than $2°$, $\sin\left[\frac{\delta x}{2}\right] = \left[\frac{\delta x}{2}\right]$, correct to 3 significant figures

i.e. $$\frac{\sin\left[\frac{\delta x}{2}\right]}{\left[\frac{\delta x}{2}\right]} = 1$$

Hence $$f'(x) = \lim_{\delta x \to 0} \left\{ \cos\left[x + \frac{\delta x}{2}\right] \frac{\sin\left[\frac{\delta x}{2}\right]}{\left[\frac{\delta x}{2}\right]} \right\}$$

i.e. $\quad f'(x) = \cos x$
Hence when $y = f(x) = \sin x$, $f'(x) = \cos x$

3. Differential coefficient of cos x

Let $f(x) = \cos x$
then $f(x + \delta x) = \cos(x + \delta x)$

By definition, $$f'(x) = \lim_{\delta x \to 0} \left\{ \frac{f(x + \delta x) - f(x)}{\delta x} \right\}$$

$$= \lim_{\delta x \to 0} \left\{ \frac{\cos(x + \delta x) - \cos x}{\delta x} \right\}$$

Now $\cos x - \cos y = -2 \sin\left[\frac{x + y}{2}\right] \sin\left[\frac{x - y}{2}\right]$ (see Appendix B).

Hence $$\cos(x + \delta x) - \cos x = -2 \sin\left[\frac{(x + \delta x) + x}{2}\right] \sin\left[\frac{(x + \delta x) - x}{2}\right]$$

$$= -2 \sin\left[x + \frac{\delta x}{2}\right] \sin\left[\frac{\delta x}{2}\right]$$

$$\frac{\cos(x+\delta x)-\cos x}{\delta x}=\frac{-2\sin\left[x+\frac{\delta x}{2}\right]\sin\left[\frac{\delta x}{2}\right]}{\delta x}$$

$$=-\sin\left[x+\frac{\delta x}{2}\right]\frac{\sin\left[\frac{\delta x}{2}\right]}{\left[\frac{\delta x}{2}\right]}$$

Hence $f'(x)=\lim_{\delta x\to 0}\left\{-\sin\left[x+\frac{\delta x}{2}\right]\frac{\sin\left[\frac{\delta x}{2}\right]}{\left[\frac{\delta x}{2}\right]}\right\}$

$=-(\sin x)$, since in the limit, $\dfrac{\sin\left[\frac{\delta x}{2}\right]}{\left[\frac{\delta x}{2}\right]}=1$

i.e. $\quad f'(x)=-\sin x$

Hence when $y=f(x)=\cos x$, $f'(x)=\sin x$

4. Differential coefficient of e^{ax}

Let $f(x)=e^{ax}$
then $f(x+\delta x)=e^{a(x+\delta x)}$

By definition, $f'(x)=\lim_{\delta x\to 0}\left\{\dfrac{f(x+\delta x)-f(x)}{\delta x}\right\}$

$$=\lim_{\delta x\to 0}\left\{\frac{e^{a(x+\delta x)}-e^{ax}}{\delta x}\right\}$$

$e^{a(x+\delta x)}-e^{ax}=e^{ax}(e^{a\delta x}-1)$

Since $e^{x}=1+x+\dfrac{x^2}{2!}+\dfrac{x^3}{3!}+\ldots$

then $e^{a\delta x}=1+(a\delta x)+\dfrac{(a\delta x)^2}{2!}+\dfrac{(a\delta x)^3}{3!}+\ldots$

Therefore $e^{ax}(e^{a\delta x}-1)=e^{ax}\left(1+a\delta x+\dfrac{(a\delta x)^2}{2!}+\ldots-1\right)$

$$=e^{ax}\left(a\delta x+\frac{(a\delta x)^2}{2!}+\ldots\right)$$

$$\frac{e^{a(x+\delta x)}-e^{ax}}{\delta x}=e^{ax}\left(a+\frac{a^2\delta x}{2!}+\ldots\right)$$

Hence $f'(x) = \lim_{\delta x \to 0} \left\{ e^{ax}\left(a + \frac{a^2 \delta x}{2!} + \ldots \right) \right\}$

$= (e^{ax})(a)$

i.e. $f'(x) = ae^{ax}$

Hence when $y = f(x) = e^{ax}$, $f'(x) = ae^{ax}$

5. Differential coefficient of ln *ax*

Let $f(x) = \ln ax$
then $f(x + \delta x) = \ln a(x + \delta x)$

By definition, $f'(x) = \lim_{\delta x \to 0} \left\{ \frac{f(x + \delta x) - f(x)}{\delta x} \right\}$

$= \lim_{\delta x \to 0} \left\{ \frac{\ln a(x + \delta x) - \ln ax}{\delta x} \right\}$

$$\ln a(x + \delta x) - \ln ax = \ln ax\left(1 + \frac{\delta x}{x}\right) - \ln ax$$

$$= \ln ax + \ln\left(1 + \frac{\delta x}{x}\right) - \ln ax$$

$$= \ln\left(1 + \frac{\delta x}{x}\right)$$

Now $\ln(1 + x) = x - \frac{x^2}{2} + \frac{x^3}{3} - \ldots$

Therefore $\ln\left(1 + \frac{\delta x}{x}\right) = \frac{\delta x}{x} - \frac{\left(\frac{\delta x}{x}\right)^2}{2} + \frac{\left(\frac{\delta x}{x}\right)^3}{3} - \ldots$

$$\frac{\ln a(x + \delta x) - \ln ax}{\delta x} = \frac{1}{x} - \frac{\delta x}{2x^2} + \frac{\delta x^2}{3x^3} - \ldots$$

Hence $f'(x) = \lim_{\delta x \to 0} \left\{ \frac{\ln a(x + \delta x) - \ln ax}{\delta x} \right\}$

$= \lim_{\delta x \to 0} \left\{ \frac{1}{x} - \frac{\delta x}{2x^2} + \frac{\delta x^2}{3x^3} - \ldots \right\}$

i.e. $f'(x) = \frac{1}{x}$

Hence when $y = f(x) = \ln ax$, $f'(x) = \frac{1}{x}$

Index